OFF GRID SURVIVAL PROJECTS BIBLE

The Complete DIY Guide to Achieving 100% Self-Sufficiency

Field-Tested Off-Grid Practices to Protect Your Family in Any Crisis

Lucas T. Fremont

Copyright © 2024 by Lucas T. Fremont

All rights reserved. No part of this book may be used or reproduced by any means, graphic, electronic, or mechanical, including photocopying, recording, taping, or by any information storage retrieval system, without the written permission of the publisher except in the case of brief quotations embodied in critical articles and reviews.

TABLE OF CONTENTS

Introduction 5
1. Off-Grid Water Purification 7
2. Emergency Water Storage 19
3. Essential Food Preservation Techniques 31
4. Budget-Friendly Stockpiling 48
5. Mastering the Great Outdoors 57
6. Essential Skills for Outdoor Exploration 66
7. Emergency Medical Skills for Survival 70
8. Emergency Sanitation 75
9. Outdoor Cooking 79
10. Solar Power 88
11. Self-Defense 97
12. Home Security Projects 106
13. Emergency Communication Techniques 115
14. Fundamental Knots for Every Situation 120
15. Survival Gardening 127
16. Hunting, Trapping, and Fishing 135
17. Sustainable Harvest 142
18. Firearms and Ammunition 147
19. Homesteading Projects 155
20. Crafting a Survival Kit 170
21. Emergency Medical Care 174
22. Vehicle Maintenance 180
23. Psychological Readiness for Off-Grid Survivalists 185
24. Survival Mindset & Mental Resilience 189
25. Survival Strategies for Natural Disasters 193
26. Herbal Medicine Essentials 197
27. Tools for Off-Grid Survival 202
28. Building a Community 209
29. Off-Grid Energy Solutions 212
Conclusions 219

Welcome, Off-Grid Enthusiast!

Thank you for picking up *Off Grid Survival Projects Bible*. You're not just reading a book—you're stepping into a journey toward true self-sufficiency and resilience for any crisis.

To make your journey even more impactful, I've prepared **exclusive bonus content** packed with practical tools and strategies to level up your off-grid lifestyle. These resources are free to you—simply scan the QR code below to access them instantly.

What's Inside?

1. **Seasonal Off-Grid Preparedness Guide**
   Stay prepared all year with tips tailored to every season.
2. **Renewable Energy Optimization Checklist**
   Maximize the efficiency of your off-grid power systems with this detailed guide.
3. **Secret Storage Solutions**
   Protect your valuables and supplies with stealthy, DIY storage ideas.
4. **Prepping Master Checklist**
   A comprehensive guide to organize and prioritize every aspect of your off-grid setup.

How to Access Your Bonuses

1. Open your smartphone or tablet camera.
2. Point it to the QR code below.
3. Click the link that pops up to unlock your exclusive content!

These tools are here to support your journey to independence. Let's build a future where you and your family thrive off the grid.

Stay resourceful,
Lucas T. Fremont
Author of *Off Grid Survival Projects Bible*

Introduction

In a world that's unpredictable, the comforts we're used to can disappear in an instant. Survival isn't just about luck—it requires knowledge, preparation, and the determination to keep going, no matter what. Whether it's a natural disaster, the collapse of society, or an unexpected emergency, those who are ready to adapt and live off the grid will have the best chance of surviving. This guide provides you with the essential skills and strategies to face these situations with confidence and resilience.

Starting with the basics is crucial—knowing how to secure clean drinking water, preserve food, and stockpile enough supplies for your family. Clean water becomes vital when usual sources aren't available, and effective food preservation methods can stretch your resources when fresh supplies are scarce. You'll also learn how to build a stockpile of lifesaving essentials without breaking the bank.

Beyond food and water, wilderness survival skills are equally important. When things go south, the ability to navigate the outdoors and utilize natural resources becomes essential. You'll explore wilderness navigation, bushcraft techniques, foraging for edible plants, and hunting, trapping, and fishing for sustainable food sources. This guide will teach you how to turn nature from a threat into an ally.

But survival isn't just about what you do outside. Protecting yourself and your home is equally important. You'll gain tactical knowledge on how to defend against threats—whether from intruders or desperate individuals searching for supplies. You'll also learn how to set up emergency communication plans to stay connected with loved ones and emergency services, even if conventional communication systems fail.

Another vital aspect of preparedness is understanding alternative energy sources. When the grid goes down, solar power can be your best friend. This guide covers DIY solar power systems and other renewable energy options, ensuring your home remains powered during outages. You'll also learn about emergency sanitation methods to maintain hygiene and prevent disease when regular systems break down.

Your survival kit deserves careful consideration as well. From assembling a bug-out bag filled with essential items for a quick escape, to basic vehicle maintenance tips that ensure your transportation holds up in a crisis—these chapters guide you through staying prepared for sudden evacuations or long-term emergencies. Mental resilience is just as critical as physical preparation; techniques for staying calm under pressure and managing stress will help you navigate the emotional challenges of any crisis.

This guide also emphasizes the power of community. Survival doesn't always mean going it alone; networking, collaborating, and creating mutual aid systems can provide extra strength during a crisis, whether you're in rural homesteads or urban environments.

Are you aiming to master sustainable homesteading skills? Interested in cultivating a survival garden or exploring the medicinal power of herbs when modern medicine isn't available? Each chapter delves into the skills needed for self-sufficient living, with clear, actionable steps to prepare you for the unknown. From emergency first aid to advanced wilderness medical care, you'll be equipped to manage injuries when professional help is out of reach.

This book is more than just a collection of survival tips—it's your comprehensive blueprint for sustainable and secure living in times of societal disruption. It covers every aspect, from essential tools and gear to broader self-reliance strategies that go beyond basic survival. As you turn each page, you'll gain the confidence that comes with being thoroughly prepared.

The future is uncertain, but with this guide, you'll face it without fear. Whether preparing for short-term emergencies or planning for long-term off-grid living, the knowledge you gain will empower you to protect your loved ones and thrive against all odds. Consider this your essential resource—one that transforms preparedness from a mere precaution into a lifestyle that ensures continuity and peace of mind, no matter what challenges life throws your way.

1. Off-Grid Water Purification

When living off the grid or in survival situations, having safe water is essential. It's crucial to understand what contaminants may be in your water and how to remove them effectively. Let's explore how to purify water when you're off-grid.

UNDERSTANDING WATER CONTAMINANTS

Water contaminants come in various forms—biological, chemical, and physical impurities. Each type poses different health risks and requires specific purification methods to ensure safe drinking water.

Biological Contaminants

Biological contaminants include bacteria, viruses, and parasites found in water. These tiny organisms can cause serious illnesses such as cholera, typhoid fever, and giardiasis. Effective purification methods are essential to eliminate these harmful pathogens and ensure water safety.

Chemical Contaminants

Chemical contaminants consist of substances like heavy metals, pesticides, and industrial pollutants. These can enter water sources through agricultural runoff, industrial waste, or natural geological processes. Specialized filtration or treatment techniques are required to remove these chemical impurities effectively.

Physical Contaminants

Physical contaminants are visible particles or sediments, such as sand, silt, or debris, present in the water. While they may not directly harm health, their presence often indicates potential biological or chemical contamination. Removing these physical impurities typically involves sediment filtration or clarification methods.

Assessing Risks in Different Water Sources

Each water source comes with its own set of contamination risks, making it essential to assess and treat each one appropriately.

Surface Water

Surface water from rivers, streams, and ponds is susceptible to contamination from animal waste, agricultural runoff, and urban pollution. To make this water safe for drinking, treatments like boiling, chemical disinfection, or filtration are often necessary.

Groundwater

Groundwater from wells and boreholes may contain naturally occurring contaminants such as arsenic or fluoride, as well as pollutants from nearby farms or factories. Regular testing and proper filtration systems are crucial to maintaining safe groundwater.

Rainwater

Rainwater is generally clean, but if collected from rooftops or catchment systems, it can pick up debris, bird droppings, or chemical residues. Proper filtration and disinfection methods are essential before using rainwater for drinking purposes.

By identifying the potential risks associated with different water sources and applying effective purification techniques, we can ensure a reliable supply of clean and safe drinking water, even in challenging situations.

Sustainable Water Filtration Techniques

Looking to purify water naturally and sustainably? Sand and gravel filters, along with charcoal-based filtration, are among the best methods available.

Sand and Gravel Filters

These filters work by physically trapping particles as water passes through layers of sand and gravel of different sizes, effectively removing many large impurities.

Materials Needed:

- A container with a hole at the bottom, such as a bucket, barrel, or any vessel that can hold water and has a drainage hole.
- Fine sand for the bottom layer.
- Coarse gravel for the top layer.
- Medium gravel to go between the fine sand and coarse gravel.
- A piece of cloth or fine mesh to cover the layers and keep the sand separate.

Layering Process:

1. **Start with Fine Sand:** Place the fine sand at the bottom of your container to trap smaller particles.
2. **Add Medium Gravel:** Next, add the medium gravel layer.
3. **Top with Coarse Gravel:** Finish by adding the coarse gravel on top to catch larger particles.

Adding Cloth or Mesh:

Lay a piece of cloth or mesh on top of the coarse gravel to prevent sand from mixing with the filtered water.

Pouring Water:

Slowly pour water over the layered materials. As the water flows through the layers, it will leave impurities behind.

Maintenance Tips:

- **Regular Cleaning:** Occasionally remove the top layers and rinse them thoroughly.
- **Replace Cloth/Mesh:** If the cloth or mesh becomes clogged, replace it to maintain smooth operation.
- **Refresh Gravel:** Replace the gravel if it becomes too compacted to ensure free water flow.

Charcoal-Based Filtration

Charcoal filtration, also known as activated carbon filtration, is excellent for removing chemicals and odors due to its high adsorption capacity.

Preparing the Charcoal:

- **Crush Charcoal:** If you have charcoal briquettes or lumps, crush them with a hammer or mortar and pestle.
- **Use Activated Charcoal Granules:** Alternatively, you can purchase activated charcoal granules, which are processed for high adsorption efficiency.

Assembling the Filter:

1. **Select a Container:** Choose a suitable container like a bucket or an empty soda bottle.
2. **Add a Fine Sand Layer:** Place a layer of fine sand at the bottom to prevent the charcoal from escaping.
3. **Add Activated Charcoal:** Evenly distribute a layer of activated charcoal on top of the sand.
4. **Top with Another Sand Layer:** Add another layer of fine sand above the charcoal for added stability.

Pouring Water:

Slowly pour water into your filter setup. As the water passes through the layers, the activated charcoal will absorb impurities like chemicals and odors.

Maintenance Tips:

- **Replace Charcoal:** Swap out the charcoal periodically as it becomes saturated with impurities.

- **Clean the Filter:** Regularly rinse the sand and charcoal layers to remove trapped pollutants.
- **Check Sand Layers:** Look for signs of clogging or compaction and replace the sand if necessary to ensure proper flow and filtration.

And there you have it—a straightforward guide to purifying water naturally! Enjoy cleaner water with these easy, sustainable methods.

DIY WATER PURIFICATION SYSTEMS

Creating your own water purification systems is practical, effective, and accessible to everyone. From constructing a bio-sand filter to assembling a solar water disinfection unit, these methods are both affordable and efficient. Let's dive into the details:

Building a Bio-Sand Filter

A bio-sand filter is particularly useful in areas where clean water is scarce. This method leverages natural processes to remove contaminants, making it reliable for both emergency situations and daily use.

Gathering Materials and Preparing the Container:

Begin by selecting a sturdy container, such as a plastic drum or large bucket, that is durable and has a lid. Thoroughly clean and disinfect the container to eliminate any contaminants. If needed, drill holes at the bottom to allow for drainage.

Assembling the Filter Layers:

First, place a layer of clean gravel at the bottom of the container. This layer supports the sand and prevents clogging. Next, add a layer of fine sand on top of the gravel; the sand will trap particles and pathogens. Optionally, you can add a layer of activated charcoal between the gravel and sand to help remove odors, tastes, and some chemicals.

Setting Up the PVC Pipe and Testing the Filter:

Position a PVC pipe vertically inside the container, ensuring it extends above the sand layer, and seal around the pipe's base to prevent water from bypassing the filter. Before using the filter for drinking water, test it by filling it with untreated water to ensure the filtered water is clear and tastes clean.

Maintaining the Filter:

Regularly clean the filter by replacing the top layer of sand, and inspect the system for any damage, making necessary repairs to keep it functioning efficiently.

Assembling a Solar Water Disinfection Unit (SODIS)

A solar water disinfection unit offers an eco-friendly water purification method using sunlight to eliminate harmful microorganisms.

Selecting and Preparing the Container:

Choose a transparent container, such as a PET bottle or glass jar, that allows sunlight to penetrate and purify the water. Fill the container with water from natural sources like rivers or streams, ensuring it is filled to the brim, and tightly seal the container with a cap to prevent contamination and evaporation.

Sunlight Exposure and Disinfection:

Place the container in direct sunlight on a flat surface, such as a roof or field, where it can receive maximum sunlight throughout the day. Allow the container to sit in the sunlight for at least 6 hours (or longer), as the UV rays work to kill harmful microorganisms. Periodically check the container to ensure it remains undisturbed.

Using and Repeating the Process:

After the recommended time, verify that the water appears clear and safe to drink. Transfer the purified water into sanitized containers for storage and use it as you would regular drinking water. If relying on untreated water sources or during emergencies, repeat this process daily to ensure a continuous supply of clean water.

CHEMICAL WATER TREATMENT: IODINE & CHLORINE

Chemical water purification is a really handy method for cleaning up water, especially in emergencies or when clean water is hard to come by. Two common chemicals doing most of the work here are iodine & chlorine. Each of them has its own perks & downsides. Now let's dig into the nitty-gritty.

Purifying Water with Iodine

Iodine is a pretty versatile chemical that zaps a lot of nasty stuff in the water, like bacteria, viruses, and those pesky protozoa. It comes in different forms—tablets, crystals, tinctures—super handy for various needs. Here's a quick guide on how to use iodine:

Prepare an Iodine Solution

Dissolve iodine tablets or maybe crystals in water to make a concentrated solution. Follow the instructions on the pack for the right dosage. Usually, it's between 8 to 16 milligrams per liter of water.

Add to Water

Pour the iodine solution into the water you want to treat. Mix it up well so it gets evenly spread out.

Wait for Contact Time

Let the iodine sit and mingle with the water for enough time to disinfect it properly. The recommended contact time can vary with water temperature and turbidity, but it's typically from 30 minutes to 4 hours.

Filter (Optional)

If your water looks murky with sediments or particles, think about filtering it before adding iodine. This can make purification more effective.

Test for Residual Iodine

After waiting (the contact time), check the iodine levels using test strips. Make sure it's within safe consumption levels, which is usually below 2 milligrams per liter.

Aerate (Optional)

To make iodine-treated water taste better, aerate it by pouring it back and forth between clean containers or letting it sit out in the open air for a bit before drinking.

CHLORINE USE IN WATER DISINFECTION

Chlorine is a super common disinfectant that knocks out bacteria, viruses & other baddies in water. You can get it as chlorine tablets, bleach, or even in liquid forms. Here's how to use chlorine for water purification:

Preparing a Chlorine Solution:

First things first, you got to dilute the bleach or tablets in water to make a solution with the right concentration. Usually, it's about 2 to 4 drops of bleach per liter of water or one tablet per quart (that's a liter too).

Thorough Mixing:

Next up, stir that chlorine solution into the water real good. You want to make sure it's all mixed up for even disinfection.

Allowing for Contact Time:

Now let the chlorine do its thing. It needs time to work, typically around 30 minutes to 4 hours. Be patient!

Optional Filtration:

If your water looks cloudy or has stuff floating in it, think about pre-filtering it. Getting rid of sediment beforehand helps the chlorine work better.

Testing for Residual Chlorine:

Grab some test strips and check for residual chlorine in your treated water. The safe range is usually between 0.2 to 4 milligrams per liter.

Optional Aeration:

Just like with iodine-treated water, aerating chlorine-treated water can help get rid of that chlorine smell & make it taste better.

Advantages and Disadvantages of Chemical Treatment:

Advantages:
- It's effective against lots of nasty microorganisms.
- Convenient and easy to carry around, especially in emergencies.
- Pretty cheap compared to other methods.
- No need for special gear.

Disadvantages:
- Can leave a weird taste or smell.
- Some folks might react badly to the chemicals.
- Effectiveness might drop if the water is too cold, has a funky pH, or is too murky.
- Might need extra steps to get rid of leftover chemicals.

BOILING AND WATER DISTILLATION METHODS

Making sure your water is safe? That's super important, especially when you're off the grid. There are two great ways to do this: boiling and distillation. Let's take a closer look at each method.

Steps for Successful Boiling

Boiling is easy and dependable. By heating water, you can eliminate harmful things like bacteria, viruses, and parasites. Here's how you do it:

Selecting the Appropriate Pot

Pick a clean pot or kettle made from non-reactive material like stainless steel or glass. Avoid aluminum, as it can make your water taste funny.

Filling Your Pot

Fill the pot with untreated water but leave some space at the top to prevent spilling over. If possible, use water from cleaner sources like tap water, spring water, or rainwater.

Bringing the Water to Boil

Place the pot on a stable heat source—maybe a stove or campfire. Bring the water to a rolling boil and keep it boiling for a full minute. If you're up high in the mountains, boil it longer to ensure all germs are killed.

Cooling Boiled Water

Let the boiled water cool naturally before using it. To cool it faster, place the pot in cold water or let it sit uncovered until it's safe to touch.

Storing Water Safely

After cooling, transfer the water to clean containers with tightly sealed lids. This keeps it safe and prevents recontamination.

DIY Process for Water Distillation

Distillation is another solid way to purify your water. It's especially effective for removing heavy metals, salts, and chemicals. Here's how you can set up your own simple distillation system:

Setting Up a Simple Distillation System

You'll need a heat-resistant pot or kettle and a condensation coil (usually a copper or stainless-steel tube). Attach this coil tightly to the lid of your container so no steam leaks out.

Filling the Distillation Container

Pour untreated water into your container, but leave some space at the top for steam to prevent overflow.

Heating the Water for Distillation

Place your container on a heat source and bring it to a boil. Steam will rise and enter the condensation coil.

Collecting Distilled Water

Position the coil away from the heat so the steam cools down and condenses back into liquid as it travels through. The distilled water will drip into another clean container.

Cooling and Storing Distilled Water

Let the distilled water cool to room temperature before transferring it to sanitized containers. Seal them tightly to maintain purity.

Benefits of Boiling and Distillation

Both boiling and distillation offer significant advantages for purifying water:

- **Accessibility:** Simple equipment and a heat source—anyone can do this!
- **Effectiveness:** Eliminates harmful microorganisms and contaminants.
- **Reliability:** Whether at home, in nature, or during emergencies, these methods reliably provide clean drinking water.

PURIFYING WATER WITH ULTRAVIOLET LIGHT

Ultraviolet (UV) light purification is a neat way to disinfect water. It really zaps harmful microorganisms, making sure your drinking water is safe. So, let's dive into how UV purification works and how solar UV disinfection methods can be easily put to use.

The Mechanics of UV Purification:

UV purification uses the mighty power of ultraviolet light to mess with the DNA of bacteria, viruses, and other nasty things. Here's a simple breakdown:

Exposure to UV Light:

Water passes through a chamber or reactor that contains a UV lamp. This lamp beams out UV-C light (with wavelengths between 200 and 300 nanometers), which targets the DNA of microorganisms.

Disruption of DNA:

When these bugs are hit with UV-C light, their DNA absorbs the radiation, damaging their genetic material. They can't cause trouble anymore because they can't replicate.

Ensuring Effective Disinfection:

By disrupting the DNA of these pathogens, UV light makes them harmless and removes the risk of getting sick from waterborne illnesses. Plus, it doesn't use chemicals or alter the taste or smell of water. Super safe and efficient!

Solar-Powered UV Disinfection Techniques:

Now, about solar UV disinfection methods—they use good old sunlight to purify water. Perfect for off-grid spots or emergencies when there's no electricity. Here are two common ways:

Solar Disinfection (SODIS) Process:

SODIS is super easy and cheap! You just need PET bottles and sunlight.

- **Filling the Bottles:** Fill clear PET bottles with untreated water, leaving some air space at the top.
- **Sunlight Exposure:** Place these filled bottles in direct sunlight for 6 hours on a sunny day—or 2 days if it's cloudy.
- **Safety Measures:** Ensure the bottles are on a reflective surface and aren't shaded by trees or buildings so they get full sun exposure.

Enhanced Solar Water Disinfection (SODIS+):

SODIS+ is like SODIS but better because it uses reflective panels to boost UV exposure.

- **Building Reflective Panels:** Use aluminum foil or shiny materials to make reflective panels and set them up to direct more sunlight onto the water-filled bottles.
- **Following SODIS Guidelines:** Fill PET bottles with water, expose them to sunlight as before, and adhere to the SODIS times.

Advantages of UV and Solar UV Disinfection:

Both UV purification and solar UV disinfection have great perks for treating water:

High Level of Efficiency:

UV light is highly effective at neutralizing many pathogens.

Chemical-Free Approach:

No chemicals involved—totally safe and kind to Mother Earth.

Easy Accessibility:

Solar methods can be done with simple materials you already have, making them great for remote areas with limited resources.

CUTTING-EDGE FILTRATION TECHNOLOGIES

Advanced filtration technologies often play a huge role in ensuring we get clean, drinkable water. Let's check out two awesome methods: ceramic filters and reverse osmosis systems, especially for off-grid living. These technologies provide practical ways to purify water, even in tough or remote places.

Ceramic Water Filters

Ceramic filters do a great job of removing nasties from water, making them popular with both households and outdoor enthusiasts. Here's how they work and how you can use them:

How Filtration Works:

Ceramic filters are made from a porous ceramic material with tiny holes that trap unwanted stuff. Contaminants like bacteria, protozoa, and dirt get blocked, while clean, tasty water flows through.

Installation and Usage:

- **Choose Your Filter:** Pick a ceramic filter based on your needs—whether it's for home use or outdoor adventures.
- **Prepare the Filter:** Soak the filter in clean water to remove any dust or particles.
- **Attach to a Container:** Ceramic filters can be attached to various water containers, like pitchers or gravity-fed setups.

- **Purify Your Water:** Pour your untreated water into the container with the filter and let it work. You'll end up with clean, safe drinking water.

Maintenance Tips:
- **Regular Cleaning:** Clean your ceramic filter regularly with a soft brush and water to remove any build-up.
- **Timely Replacement:** Follow the manufacturer's guidelines for replacing the filter cartridge to ensure it keeps working effectively.

Reverse Osmosis Systems for Remote Areas

Reverse osmosis (RO) systems are super effective at purifying water, even in off-grid locations where access to clean water can be tricky. Here's how you can set one up for off-the-grid living:

Key Components of the System:
- **RO Membrane:** The core of the system. The RO membrane filters out contaminants through osmosis, providing clean water.
- **Pre-filters:** These remove larger particles and sediment before the water reaches the RO membrane, helping it last longer.
- **Post-filters:** Post-filters provide a final polish by removing any remaining impurities and improving taste.

Installation Process:
- **Select the Right System:** Choose an RO system that suits your needs, considering your water usage and available space.
- **Assemble the Components:** Follow the manufacturer's instructions to assemble the pre-filters, RO membrane, and post-filters.
- **Connect to a Water Source:** Hook up the system to a water source like a well, river, or rainwater tank using the appropriate tubing and fittings.

Operating the System:
- **Activate the System:** Switch on the RO system and let it run to produce clean water.
- **Monitor Functionality:** Keep an eye on its performance—watch for leaks or any glitches.

Maintenance Guidelines:
- **Filter Replacement:** Change out pre-filters, the RO membrane, and post-filters according to the manufacturer's schedule to maintain water quality.
- **System Sanitization:** Regularly clean the system to prevent bacterial growth and ensure it continues to function well.

PROJECT 1: BUILD YOUR OWN RAINWATER COLLECTION SYSTEM

The best way to ensure your water is fresh is to catch rain and then clean it up – it's already been through a filtration process, so it is almost certainly safer than anything you find running across the lands or underground. Creating your own rainwater catchment system isn't particularly difficult, but it can be a bit time consuming. This project details what you need, then walks you through the steps to make your own rainwater system.

Materials and Preparation

Before you start, you will need the following materials:

- Water storage barrels, usually between 30 and 55 gallons
- 1 inch spigot with a ¾ inch pipe thread
- 1.75-inch x 0.75 inch coupling
- 1.75-inch x 0.75 inch bushing
- 1.75-inch lock nut
- 1.75-inch pipe threat with hose adapter
- Metal washers
- Teflon thread tape
- Silicone caulk
- S-shaped downspout elbow
- Aluminum window screen
- Concrete blocks

Steps

You can start by modifying the barrel.

1. Use a drill to add a hole about a foot from the bottom of the barrel. The hole should be roughly 0.75 inches so that the spigot fits snuggly in it. If you have a different sized spigot, you will need to make the hole the proper size for it.

2. Add caulk around the hole you just drilled, both inside and outside.
3. Insert the spigot and the coupling into the hole, then add the Teflon tape to tighten the seal.
4. Add the overflow valve.
    a. Drill a second 0.75 inch hole a little higher, ensuring you can add a bucket or other tool under it.
    b. Add caulk around the second hole, both on the outside and the inside.
    c. Add a washer on the outside, and another on the inside, and apply the tape to them.
5. Cut a hole in the top of the barrel that is large enough for the downspout elbow to fit.

Now you can add an outdoor hose to the barrel to run water if needed.

Common Types of Water Barrels

Repeat this until you have as many water barrels as you want for storing water.

Once your barrel (or barrels) are ready, you can prepare the area to redirect one of the downspouts of your home. You will need to complete the following tasks first if the downspout you redirect is not on concrete or other flat, artificial surface.

1. Create a level area near one of the downspouts on your home. You are going to adjust the downspout so that it delivers water into your new system, so you need a level place to place the barrel to catch the water.

2. Add pea gravel to the newly leveled area. This will act as a drainage system so that water does not damage the foundations of your house.
3. Ade concrete blocks on the gravel. Make sure to evenly space them sideways so that they make a raised platform.

Finally, you need to redirect the water and set up the system.

1. Remove the lower part of the downspout and add the downspout elbow.
2. Add the lower end of the elbow to the hole in the top of the barrel, then add some of the metal screen to protect the area around the hole. This should keep debris, bugs, and other things from getting into your collected water.
3. Install a filter screen over the gutters to keep large debris from getting into the gutters and going down the downspout. This will have the added benefit of simplifying the process of cleaning your gutters.

The following is an image of a system with three barrels, which is a common configuration for a home system.

PROJECT 2: CREATE A DIY WATER FILTER

Water filters are surprisingly easy to make – it's a popular project for kids and they are made on survival shows. This project will walk you through what you need to do to create your own water filter.

Materials and Preparation

Before you start, you will need the following materials:

- A plastic water bottle that has a cap
- Knife, preferably a crafty knife, exacto knife, or a pocket knife
- Hammer
- Sterilized nail, or a nail that isn't rusted or dirty
- Coffee filter
- Large cup
- Container for the filtered water. Activated charcoal
- Gravel
- Sand (not colored sand)

Steps

You can start by modifying the barrel.

1. Use a drill to add a hole about a foot from the bottom of the barrel. The hole should be roughly 0.75 inches so that the spigot fits snuggly in it. If you have a different sized spigot, you will need to make the hole the proper size for it.
2. Cut roughly an inch off of the bottom of the bottle. It will now have a large opening in the bottle.
3. Place the nail in the center of the bottle cap, then hammer it so that the nail goes through the cap. You don't need to drive the nail through the cap; you just need to make the hole so that it goes all the way through the cap.

4. Place the coffee filter over the top of the bottle, then screw the cap onto the bottle. There will now be a filter between the bottle and the cap.
5. Turn the bottle upside-down, then place the cap in the container for catching the water. This could be a mug, large cup, or a jar.
6. Add the activated charcoal into the bottle, filling it roughly a third of the way up the bottle.
7. Add the sand to the bottle, filling it another third of the way to the top.
8. Add the gravel to fill the rest of the bottle. You should now have three layers in the bottle, with the gravel being at the top.

Now you can pour water into the filter, and the three layers will remove most of the debris and particles from the water. You will still need to sterilize the water before you can drink it, but you will have at least gotten most of the debris and large particles out of the water.

2. Emergency Water Storage

INTRODUCTION TO STORING WATER FOR EMERGENCIES

When it comes to being prepared, water is key. It's the cornerstone of survival. As you begin your journey to understand emergency water storage, always remember: in a crisis, water becomes even more valuable than gold. Whether you're a seasoned prepper or just starting out, knowing how to store water for emergencies can be a game-changer. It can mean the difference between thriving and just getting by.

The Critical Role of Water During Emergencies

Water is the main building block of life, and this becomes especially clear during emergencies. Natural disasters might disrupt municipal water supplies, or you could find yourself stranded without access to fresh water due to unforeseen circumstances. These scenarios are varied and unpredictable. We need water for many essential activities—drinking, cooking, hygiene, and medicine. During emergencies, turning on a tap and having clean, drinkable water isn't always an option. This highlights why having a reliable water storage plan is so important (although realizing this might come a bit too late).

Determining Your Emergency Water Needs

It's generally recommended to store at least one gallon per person per day. However, this is the bare minimum. For added safety, aim for two gallons per day. Consider your family size, pets, and the climate—hot conditions could increase your need for more water. For a family of four over two weeks, you'd need about 56 gallons at minimum. But why stop there? Storing enough for a month or more gives you a bigger safety net and peace of mind.

SELECTING THE RIGHT WATER STORAGE METHOD

Selecting the right method to store your water means finding the balance between space, budget, and your household's needs.

Water Storage Container Options

Commercial Water Storage Containers: These are designed specifically for storing water and are made from food-grade plastic. They come in sizes ranging from 5-gallon jugs to 55-gallon barrels and often include UV protection and durable construction to prevent leaks.

Water Bricks: These are compact and stackable, making them great if you're tight on space. Each brick typically holds around 3.5 gallons, and their design makes them easy to stack.

IBC Totes: These containers hold large amounts of water (275 gallons or more). They're good for long-term storage and can be sourced from industrial suppliers.

Evaluating the Pros and Cons of Water Storage Options

Each method has its advantages and disadvantages:

Commercial Water Storage Containers are user-friendly and reliable but can be pricey when aiming for extensive storage.

Water Bricks are perfect if you have limited space since they're modular and easy to handle. However, their smaller capacity means you'll need quite a few bricks.

IBC Totes stand out due to their large capacity and durability, making them economical for bulk storage, but they're difficult to move once filled.

Choosing the best storage solution is crucial when preparing. Think about your available space and how much accessible water you'll need during an emergency. Always remember: the goal is not only to survive but to thrive even in challenging times.

Effective Water Collection Methods

Knowing how to collect water properly is important when focusing on self-reliance and resilience.

Rainwater Harvesting Techniques

Harvesting rainwater is a vital step toward self-sufficiency during times of scarcity.

Understanding How to Harvest Rainwater

This involves catching and storing rainwater that lands on your property instead of letting it wash away down drains.

Components Needed:

Catchment Surface: The area where rain will be collected (like your roof). Gutters/Downspouts: Direct rainwater from the catchment surface. Storage Tank/Barrel: Stores the collected rainwater. Filtration System: Ensures the water is clean. Overflow Mechanism: Manages excess water during heavy rainfall. Installing Your System

Here's how:

Choose your catchment surface wisely. Attach gutters/downspouts that direct rain into storage containers/tanks. Position your tank/barrel on level ground close by. Connect funnels securely between gutter guards and tanks/barrels, avoiding leaks. Filter debris properly, checking regularly to ensure no blockages occur, maintaining optimal functionality year-round.

Safely Collecting Water from Natural Sources

Collecting safe drinking water from natural sources is essential for survival, especially in remote areas during emergencies. Having the knowledge and skills to collect and purify water can be immensely beneficial:

Assessing Sources: Before collecting, consider the suitability:

Visual inspection to check for pollution, contamination, discoloration, odors, or debris. Assess animal activity, as frequent visits or waste could affect water quality. Avoid potential pollutants from nearby settlements or industrial areas. Filtering and Purifying: Once a safe source is identified, the next steps involve making the water consumable:

Filtering: Remove suspended particles and sediment using portable filtration options. Boiling: Effective for sterilizing pathogens; bring water to a rolling boil for about a minute, and longer at high altitudes. Chemical Treatment: Chlorine dioxide tablets or iodine can disinfect water; follow manufacturer guidelines for dosages and treatment times.

Important Tips and Precautions

Boil: Boiling is a trusted method for pathogen eradication, offering a simple and preventive solution. Clean Containers: Use clean containers made from safe materials to avoid contamination. Rain Harvest Techniques: Setting up systems like rooftops and barrels for efficient collection is crucial. Use clean apparatus to avoid contamination and mesh filters to refine captured water. Rivers and Lakes: Select sites with minimal contamination. Avoid pollutants by choosing upstream sources and using midstream positions to optimize water quality. Filtration and boiling remain crucial steps for ensuring safe, drinkable water. Remember, the goal is to maximize safety and ensure a reliable supply of potable water in any situation.

WATER PRE-TREATMENT FOR LONG-TERM STORAGE

Making sure your water stays safe and fresh for long-term storage is super important, especially when you're prepping for emergencies or survival situations. Pre-treatment helps remove impurities and germs, keeping your water good to drink over time. Let's dig into the pre-treatment process. First, we'll talk about sediment filtration, and then we'll dive into chemical treatment, mainly using chlorine.

Sediment Filtration

Sediment filtration is the first step when you're getting water ready for long-term storage. It removes stuff like dirt, particles, and other bits that can make the water cloudy. Here's how you can do it right:

- **Selecting a Filtration System:** Find a good filtration system that works for you. It could be a portable water filter or maybe a gravity-fed unit.
- **Preparation:** Before you filter the water, let it sit in a container. This allows the big bits to settle at the bottom.
- **Filtration Process:** Pour the settled water through your filter system. These filters usually have layers that catch the sediment and particles.
- **Replacement of Filter Elements:** Keep an eye on your filter cartridges or media and change them regularly to keep things working smoothly.

Chemical Treatment: Chlorination

Using chlorine is a common way to clean water because it kills harmful bacteria, viruses, and parasites. Here's how you can chlorinate your water effectively:

- **Determine Chlorine Dosage:** Figure out how much chlorine you need based on how much water you're treating. Follow the guidelines!
- **Mixing Chlorine:** Mix chlorine bleach or tablets into the water to get the right concentration. Make sure you're using a clean container.
- **Distribution:** Stir or shake (carefully!) to evenly distribute the chlorine throughout the water.
- **Contact Time:** Let it sit for at least 30 minutes to an hour so the chlorine can do its job.
- **Testing:** Use those nifty test strips to make sure you've got enough chlorine in there. The levels should be safe for drinking.

Precautions and Safety Tips

A few extra tips to keep everything safe and organized:

- **Proper Storage:** Keep your treated water in clean, food-grade containers with lids that fit snugly.
- **Regular Testing:** Test your stored water every now and then to make sure there's still some residual chlorine doing its work.
- **Rotation:** Periodically switch out your stored water with fresh treated batches. Use the older stuff for things like cleaning.
- **Labeling:** Write dates and other important info on your containers so you can keep track of what's what.

PRESERVING WATER QUALITY

Keeping your water quality top-notch is super important to ensure it's safe and good enough to drink. Regular check-ups and preventive steps are a must to stop algae and harmful bacteria from taking over, ensuring your water stays clean. Let's dive into some handy tips for maintaining water quality.

Stopping Algae & Bacterial Growth

Algae and bacteria can mess up your water's quality, making it stink, taste weird, or even make you sick. So, keeping them out is crucial.

- **Sunlight Protection:** Keep your containers out of direct sunlight. Store them in shady spots or use UV-resistant ones to block sunlight.

- **Aeration:** Give the water a good stir now and then to get oxygen in there. This helps stop bad bacteria from growing and keeps the water fresh.
- **Regular Cleaning:** Clean your containers regularly. Remove any organic bits or sediment to avoid bacteria and algae breeding grounds.
- **Use of Algaecides:** Use products like copper sulfate or hydrogen peroxide to keep algae at bay, but always follow the instructions on the package.

Routine Maintenance and Inspection

Sticking to regular maintenance and inspections helps catch problems early on, ensuring your water stays top-grade.

- **Visual Inspection:** Check containers for any signs of gunk or algae. If something looks off, address it right away.
- **Cleaning Schedule:** Set up a routine for cleaning containers, filters, and distribution systems. Use mild soap and rinse well.
- **Filter Replacement:** Replace filters as recommended by the manufacturer to maintain top-notch filtration and avoid bacterial buildup.
- **Water Testing:** Test the water periodically for pH levels, chlorine, and any germs. Adjust treatments based on the results.
- **System Upkeep:** Keep an eye out for leaks or external contamination in plumbing fixtures and seals. Make sure everything is in good working order!

By focusing on these steps, your water supply will be safe for drinking and use anytime. Regular checks and upkeep are key parts of good water management.

CHOOSING STRATEGIC STORAGE LOCATIONS:

Planning where you store supplies is vital during emergencies. Picking the right spots can make all the difference in being prepared.

- **Accessibility:** Choose spots close to home. You don't want to be running far in a crisis.
- **Security:** Keep supplies safe from theft or disasters with locks, alarms, or cameras.
- **Climate Control:** Avoid places with extreme temperatures or moisture. Store items in stable environments to prevent damage.
- **Organization:** Keep areas tidy and labeled for quick access. Use shelves and cabinets wisely!

Home Storage Solutions

Having smart storage solutions at home is key for stocking up on essentials without wasting space:

- **Pantry:** Set up a pantry for non-perishables like canned goods and grains. Check expiration dates often and rotate items.
- **Emergency Kit:** Put together an emergency kit with essentials like first aid supplies, flashlights, and batteries, stored where you can reach it quickly (like a closet).
- **Water Storage:** Set aside space for water containers so you have clean drinking water ready. A filtration system can help ensure long-term safety.
- **Medicine Cabinet:** Stock it with meds, first aid supplies, and hygiene products. Regularly check expiration dates and refill as needed.

Secondary Locations: Bug-Out Sites and Vehicles

Having backup storage spots helps you be more prepared:

- **Bug-Out Sites:** Secure places like cabins where extra supplies—food, water, shelter items—are stored, ready if you need to evacuate quickly.
- **Vehicles:** Keep your car stocked with blankets, food rations, and emergency kits in designated areas for easy access on the move.
- **Mobile Storage:** Invest in backpacks filled with essentials, always within reach whether you're home or out and about.

ENSURING EMERGENCY ACCESS AND PORTABILITY

Having access to portable water solutions is super important for staying hydrated and surviving during emergencies or evacuations. Equipping yourself with the right tools and accessories can help you access stored water easily and stay hydrated while on the move.

PORTABLE WATER SOLUTIONS FOR EVACUATION

When evacuating or facing emergencies, you need lightweight, compact, and easy-to-carry portable water solutions. Here are some practical options:

- **Water Bottles:** Get durable, reusable water bottles that are great for storing and carrying water. Choose BPA-free materials and bottles with secure lids to avoid leaks.
- **Collapsible Water Containers:** Consider collapsible water containers that fold or roll up when empty. They're perfect for saving space in emergency kits or backpacks. Make sure they have a spigot for easy pouring.
- **Water Packets or Pouches:** Pack some purified water packets or pouches. They're really handy during emergencies and can fit in your pockets, bags, or emergency kits. Just ensure the water is safe to drink and meets safety standards.
- **Water Purification Tablets:** Keep water purification tablets in your emergency kit to handle uncertain water sources. These tablets effectively kill microbes, making the water safe to drink.

EQUIPMENT AND ACCESSORIES FOR RETRIEVING STORED WATER

Having the right tools and accessories is super important for accessing stored water quickly during emergencies. Here's some stuff you might want to grab:

Siphon Pump: A siphon pump comes in handy for moving water from big containers to smaller ones or bottles. Pick a pump made of food-grade materials, with a flexible hose to make things easier.

Water Filter Straw: A water filter straw is small and lightweight. You can drink straight from dirty water sources using these straws, which usually have a filter system that removes bacteria and other nasty things, so you get safe drinking water.

Water Container Spigot: Attach a spigot to your large water containers to make it simple to pour out water. Ensure the spigot doesn't leak and fits your containers properly.

Water Purification System: Consider getting a portable water purification system. It can clean large quantities of water if you need it for an extended period. Look for systems that use filters, UV light, or chemicals to ensure the water is extra clean.

Be sure to keep safe and clean drinking water available during emergencies and evacuations by having portable water solutions and tools. Opt for lightweight and compact options. Don't forget to check and refill your emergency water stash regularly, so you're always ready!

STORING CLEAN WATER

Keeping purified water safe is key to maintaining its quality and usability. Whether it's for emergencies, adventures, or just daily use, proper storage is crucial. Let's dive into the main points of safe water storage and how to spot signs of deteriorating water.

Safe Storage Practices

Container Selection:

- Pick food-grade containers specifically made for storing water. This helps prevent contamination.
- Use BPA-free containers to avoid harmful chemicals in your water.
- Choose containers with tight lids to keep air and other contaminants out.

Location:

- Store your water in a cool, dark place. Sunlight can promote algae growth.
- Keep water away from chemicals and hazardous materials, as they can contaminate it.

Cleaning and Preparation:

- Before storing water, clean the containers thoroughly. Use a mild bleach solution (1 teaspoon of bleach per gallon of water).
- Rinse the containers well to remove any residual bleach.

Filling and Sealing:

- Fill the containers with clean, safe water, leaving a bit of space at the top to allow for expansion if frozen.
- Seal them tightly to keep out air and dirt.

Rotation: Check your stored water regularly and replace it every 6 to 12 months to keep it fresh.

Indicators of Water Quality Deterioration

Even with proper storage, water quality can degrade over time. It's important to check your stored water regularly.

Appearance:

- Is it cloudy or discolored? This could indicate contamination.
- Sediment at the bottom? That's a sign the water might not be safe.

Odor: Does it smell weird? Foul or musty odors could mean bacteria or chemicals have contaminated the water.

Taste: Does it taste off? A bad taste might indicate chemicals or microorganisms have entered the water.

Container Condition: Inspect for leaks and damage. Any cracks or leaks can compromise water quality.

Microbiological Testing: For long-term storage, consider periodically testing the water for microorganisms to ensure it remains safe.

Keeping an eye on these factors helps ensure your stored purified water stays fresh and safe!

ESTABLISHING BACKUP SYSTEMS FOR WATER PURIFICATION

Redundancy is a must-have. By mixing up purification methods and preparing for multiple situations, you can ensure a steady flow of clean drinking water, even when times are tough. Use all the methods mentioned here to build your redundancy.

Let's go over these methods and compare them. This will help you decide which redundancies are best for your setup. Don't forget, you can always print this section to keep with your supplies in case you're not home during an emergency.

Diversifying Purification Methods

Boiling Water:

- **Purpose:** Boiling water is super simple and one of the best ways to kill germs. It makes water safe to drink.
- **Operation:** Bring water to a rolling boil (yep, bubbles) for at least one minute—longer if you're up in the mountains.
- **Benefits:** Boiling only requires a pot and a heat source. It's great at zapping a wide range of nasty stuff.

Filtration Systems:
- **Purpose:** Filters, like ceramic ones or reverse osmosis systems, pull out dirt, impurities, and bugs from water.
- **Operation:** Run water through the filter to catch and remove contaminants. Clean water comes out the other side.
- **Benefits:** Filters are great for long-term use and can be tailored to specific challenges in your water.

Chemical Treatment:
- **Purpose:** Chemicals like iodine or chlorine kill harmful bacteria, viruses, and parasites in water.
- **Operation:** Add the right amount of chemical to the water and wait for it to do its thing.
- **Benefits:** Chemical treatments are light and portable, making them ideal for treating large amounts of water.

UV Disinfection:
- **Purpose:** UV light disrupts microorganisms by messing up their DNA, preventing them from reproducing.
- **Operation:** Shine UV light on water with a portable UV purifier, or use sunlight in a clear bottle for solar disinfection.
- **Benefits:** UV disinfection is eco-friendly, uses no chemicals, and quickly kills pathogens.

Preparing for Multiple Scenarios

Emergency Situations:
- **Plan Ahead:** Consider emergencies that might cut off your water supply—like natural disasters or infrastructure failures.
- **Stockpile Supplies:** Keep a variety of purification methods handy. Backup options like tablets or portable filters are smart too.

Off-Grid Living:
- **Self-Sufficiency:** Embrace self-reliance by setting up multiple purification systems, ideal for off-grid living.
- **Alternative Power:** Look into solar-powered or hand-crank purifiers so you can have clean water anywhere.

Travel and Outdoor Adventures:
- **Portable Solutions:** Bring lightweight, compact purification gear when camping, hiking, or traveling where water isn't reliable.
- **Versatility:** Choose methods that work well in different locations and conditions.

With a variety of water purification options ready and proactive preparation, you're set to build a reliable drinking water system. Whether it's an emergency, off-grid living, or outdoor trips, having multiple methods ensures you always have clean water on hand.

PROJECT 3: SAVE WATER WITH A SIMPLE DIY SINK

In the event of a disaster, water conservation is going to be critical. The toilet has the greatest use of water in the home, so having a sink that uses the water tank of your toilet to conserve water is going to allow you to make your water to go further. This project details what you need, then walks you through the steps to make your own sink that will help you conserve water during an emergency.

Materials and Preparation

Before you start, you will need the following materials:

- Sink fixture that fits the top of your toilet (there are specialized fixtures you can get online or from your local hardware store)
- Extender for the sink

Steps

This project is actually fairly simple because most of the components come from a store. Your primary focus is on making sure that you have a sink that you can use should you lose access to a clean water source.

1. You need to make sure you have the right measurements for the sink before you go to purchase it.
    a. Determine which toilet tank you want to use in the case of an emergency. You may use multiple toilets if you have several people in your home to ensure that you conserve as much water as possible. You will need to measure and track those measurements for each toilet tank you plan to use.
    b. Ensure the measurements on the packaging match the toilet tank you need.
2. Remove the top from the toilet tank that you will use the most often in the event of an emergency.

3. Install the sink on top of the toilet and make sure it fits snuggly. You may need to add the extender to ensure that the sink doesn't fall into the tank.
4. Remove the sink.
5. Use the instructions that came with the sink to add the faucet to the assembly.
6. Make sure the tube from the faucet is secure.
7. Feed the tube from the faucet into the toilet tank.
8. Make sure that the sink fits snuggly, including adding the extender if needed.

Example of an Installed Water Conservation Sink

Now you have a way to save the water you've worked hard to store so that it will last longer.

PROJECT 4: LONG-TERM WATER STORAGE: BUILD YOUR OWN TANK

Depending on how much space you have, your storage unit can be a few gallon barrels or a large 6,000 gallon tank. If you have a yard that is big enough to add a large tank, that could be your best bet in the event of a major catastrophe. However, having a smaller system with several barrels is adequate for most small families as the water can be replenished over time through rain and other water saving measures.

This project details what you need, then walks you through the steps to make a smaller system of barrels for long-term storage.

Materials and Preparation

Before you start, you will need the following materials:

- Choose your storage containers (even if you have a large water storage unit, you will need smaller containers for drinking and regular use). You can use large barrels, liter drinking bottles, and unused Jerry cans (do *not* use cans that have stored any type of fuel or hazardous liquids)
- Unscented dish soap
- Unscented chlorine bleach (between 5% and 9% sodium hypochlorite)

Steps

Once you have your barrels, you need to start by cleaning them, even if they are new. You want to make sure that you aren't introducing chemicals or in any way contaminating the water you plan to store.

1. Clean the container.

a. Fill the new container up to a third full with warm water.
   b. Add the unscented dish soap.
   c. Ensure the water touches all of the inside surface of the container.
      i. If you are cleaning a barrel, lay it on the side and roll it over and over to get all of the inside of the barrel.
      ii. If you are cleaning a small container, shake it vigorously for up to a minute.
2. Pour out the water.
3. Sanitize the container.
   a. Fill the new container up to a third full with warm water.
   b. Add the unscented 1 teaspoon of bleach.
   c. Ensure the water touches all of the inside surface of the container, as done in step 1.
4. Pour out the water in a place that is environmentally safe for water that contains bleach.

Once clean, you can modify the barrels for easier retrieval of water.

5. Use a drill to add a hole about a foot from the bottom of the barrel. The hole should be roughly 0.75 inches so that the spigot fits snuggly in it. If you have a different sized spigot, you will need to make the hole the proper size for it.
6. Add caulk around the hole you just drilled, both inside and outside.
6. Insert the spigot and the coupling into the hole, then add the Teflon tape to tighten the seal.
7. Add the overflow valve.
   a. Drill a second 0.75 inch hole a little higher, ensuring you can add a bucket or other tool under it.
   b. Add caulk around the second hole, both on the outside and the inside.
   c. Add a washer on the outside, and another on the inside, and apply the tape to them.
8. Cut a hole in the top of the barrel that is large enough for the downspout elbow to fit.

   Now you can add an outdoor hose to the barrel to run water if needed.

Common Types of Water Barrels

As shown in Chapter 1, the following is an image of a system with three barrels, which is a common configuration for a home system.

PROJECT 5: MAKE YOUR OWN HAND-POWERED WATER PUMP

If you have a well, you will want to have a hand pump to simplify the process of retrieving water. This is much better for a long-term solution if you don't know when regular water service will be restored. This project details what you need, then walks you through the steps to make your own rainwater system.

Materials and Preparation

Before you start, you will need the following materials:

- 1 in PVC pipe that is 5 ft long
- 1.25 in PVC pipe that is 5 ft long
- 2 1.25 in T fittings
- 0.75 in threaded reducer
- 0.5 in threader cap
- 2 check valves (should fit the 1.25 PVC pipe)
- 1 1.25 thread fittings
- 2 O rings
- Pipe cement

You will need to make sure that the pump fits your well, so you may need to adjust the length of the pipe to fit any existing wells.

Steps

Once you have all of your supplies and the measurements for the well, you can get started making the pump.

1. Make the plunger assembly.
    a. Cut the 1 in PVC pipe so it is 3 ft long.
    b. Add shallow trenches to one end of the tube.
    c. Fit the O rings in the trenches.
    d. Insert the threaded reducer.
    e. Apply the tube cement to the reducer.
    f. Wait an hour for the cement to dry.
    g. Add the plunder to the T fitting.
2. Make the valve assembly.
    a. Cut the 1.25 in PVC pipe so you have a 3 ft length.

b. Insert the T assembly.
   c. Cut the rest of the PVC into 6 in pieces. There should be four pieces.
   d. Uses two of the pieces to make the pump handles.
   e. Add the other two pieces on both sides of the T fitting.
   f. Insert the check valves into the thread fittings.
   g. Check the arrows on the valves and make sure that you insert them facing the same direction. This is essential to ensuring that it works.
3. Insert the plunger to complete the pump.

The Finished Hand Pump

You should make sure to test the pump to make sure it works, that way you know that it is ready in case of an emergency.

PROJECT 6: TURN AIR INTO DRINKABLE WATER AT HOME

You need a few dehumidifiers for this projects, making it perhaps the easiest project in this book. This is a very short-term solution as humifies do not generate much water on their own. You will need a power source for this projects, and most of the sources in this book are enough to run these devices.

This project details what you need, then walks you through the steps to pull enough water out of the air to have water for a few days.

Materials and Preparation

Before you start, you will need the following materials:

- Several dehumidifiers (do not use units that have desiccants because the water is absorbed instead of filtered into a container)
- Water sanitation and filtration tools
- Chemical test strips

It is critical to keep your dehumidifier system clean so that mold, mildew, and other contaminants are not added to the water over time.

Steps

Once you run a dehumidifier, you can collect the water from it.

1. Thoroughly clean the system to remove de
2. Turn off the unit and remove the collection unit.

3. Use the strips to test for heavy metals and chemicals.
4. Filter and sanitize the water before you use it for drinking or ingestion.

The Water Collection Container on a Dehumidifier

An adult needs about 1 l of water a day, so you will need to be able to generate that much water through this method for every person (and pet) who needs water. This is why it is a very short-term solution. The longer you need water, the less practical this solution is. It should be a stop-gap measure until you find a more reliable water source.

3. Essential Food Preservation Techniques

It is super important to have a steady food supply. Using food preservation methods lets you stock up and keep your food fresh for long periods. This way, you can always have healthy meals whenever needed.

PRESERVING WITH CANNING: A PROVEN TECHNIQUE

Canning is a tried-and-true way of preserving food by sealing it in airtight containers, usually glass jars. It keeps the food from spoiling and makes it last much longer. With the right canning process, preppers can enjoy fruits, vegetables, meats, soups—you name it—long after they were first stored.

Getting Started with Canning

Canning is an essential skill for any prepper because it's straightforward and dependable. If you're new to canning or just want to improve your skills, knowing the key steps is essential for success.

Benefits of Canning for Preppers

- **Long-Term Preservation:** Properly canned foods can last for years, providing preppers with a reliable food supply during emergencies or shortages.
- **Versatility:** Canning allows preppers to keep a variety of foods fresh—from garden produce to homemade soups.
- **Cost-Effective:** Canning your own food can be cheaper than buying store-bought canned goods, especially if you use homegrown produce or buy in bulk.
- **Self-Reliance:** Learning how to can makes preppers more independent, reducing reliance on grocery stores and other outside sources.

Materials Needed:

Gather these materials before starting:

- **Glass canning jars with lids and bands:** Use jars made specifically for canning—they're tough and can handle high temperatures well.

- **Canning equipment:** Depending on your method (water bath or pressure canning), you'll need a boiling water canner or a pressure canner.
- **Jar lifter:** A tool that lets you safely move hot jars during the process.
- **Canning funnel:** Helps fill jars with less mess.
- **Bubble remover and headspace tool:** Removes air bubbles and ensures proper headspace in jars.
- **Clean towels/rags:** For cleaning rims and handling spills during canning.
- **Acid like lemon juice or vinegar:** Some recipes require this to ensure safe preservation, especially for low-acid foods.

Preparation Steps:

To get ready:

- Wash jars, lids, and bands thoroughly in hot, soapy water. Rinse them well to remove any soap residue.
- Inspect each jar for cracks, chips, or defects. Discard any damaged jars—they might not seal correctly.
- Prepare your recipe by gathering all ingredients and following a trusted source's steps for safety and optimal preservation.

Canning Process:

You're ready! Here's how to do the actual canning:

1. Pack prepared food into the jars using a funnel, leaving the required headspace as directed by your recipe.
2. Wipe rims clean with a damp towel to remove spills or food residue.
3. Apply lids and bands—center the lids on each jar and screw the bands fingertip-tight to ensure a proper seal.
4. Process jars according to your chosen method (water bath or pressure canning). Place filled jars into the canner and process based on the specified time and pressure.

Post-Canning Steps:

Once you're done sealing the jars:

- Carefully remove hot jars from the canner using the jar lifter. Place them on a towel-lined surface to cool.
- Let them cool undisturbed for 12-24 hours. You might even hear that pleasant "ping" sound of lids sealing during this time!
- Check seals once cooled—properly sealed jars won't flex or pop when pressed on the lid.
- Label each jar with the contents and date before storing them in a cool, dark place.

FOOD PRESERVATION THROUGH DEHYDRATION FOR LONG-TERM STORAGE

Dehydration is a time-proven way to keep foods tasty and safe. You can use it to extend the shelf life of ingredients while maintaining their flavor and nutritional value. This guide will cover the basics of dehydration and offer steps to help beginners get started.

Introduction to Dehydration

Dehydration involves removing moisture from foods, preventing bacteria, yeast, and mold from growing. By reducing the water content, microbes can't thrive, significantly extending the food's shelf life.

Benefits of Dehydration

- **Long-Term Storage:** Dehydrated foods can last for months or even years without refrigeration.
- **Nutritional Preservation:** Unlike other methods that may degrade nutrients, dehydration preserves most vitamins and minerals.
- **Space Efficiency:** Dehydrated foods are lightweight and compact, allowing for efficient storage.

Getting Started with Dehydration

Here's how to begin dehydrating your foods:

1. **Select Your Ingredients:**
   Choose fresh, ripe produce or lean meats. High-quality inputs yield the best results.
2. **Preparation:**
   Clean your ingredients. Peel, slice, or chop fruits and vegetables into uniform pieces for even drying.
3. **Pre-Treatment (Optional):**
   Some foods benefit from pre-treatment to enhance flavor, texture, or color—consider blanching, lemon juice dips, or marinating.
4. **Arrange on Dehydrator Trays:**
   Spread your prepared ingredients evenly on the trays of your food dehydrator. Good airflow is essential for efficient drying.
5. **Set Temperature and Time:**
   Refer to your dehydrator's manual for the correct temperature settings and drying times based on the specific foods you're dehydrating.
6. **Monitor Progress:**
   Keep an eye on the drying process. Rotate trays if necessary to ensure even drying.
7. **Test for Dryness:**
   Press on the dried food. It should feel dry and leathery.
8. **Cool and Store:**
   Allow your dried foods to cool completely before storing them in airtight jars or vacuum-sealed bags.

Popular Dehydrated Foods

- **Fruits:**
  Apples, bananas, strawberries, and mangoes are excellent for dehydrating and make sweet, nutritious snacks.
- **Vegetables:**
  Bell peppers, carrots, tomatoes, and onions are ideal for soups, stews, and casseroles.
- **Meats:**
  Lean beef, turkey, and chicken can be thinly sliced and dried into jerky, providing a protein-rich snack that's great on the go.

Tips for Success

- **Uniformity:**
  Cut ingredients into even pieces to ensure uniform drying.
- **Air Circulation:**
  Ensure good airflow by not overcrowding trays, allowing air to circulate freely.
- **Storage:**
  To maintain quality and extend shelf life, store dried foods in a cool, dark place away from direct sunlight.
- **Rehydration:**
  When ready to use, soak dried foods in water or add them directly to soups and stews.

FROZEN STORAGE: KEEPING PERISHABLE FOODS FRESH

Freezing is a trusty way to keep perishable foods fresh, letting you stretch the shelf life of fresh goodies and stock up on essentials. This piece dives into freezing with some handy tips and tricks.

Introduction to Freezing

So, how does freezing work? It drops the food's temperature super low, stopping the growth of little bugs (microorganisms), kind of putting them on pause. By halting bacterial activity, freezing helps keep your food in good shape.

Benefits of Freezing

- **Long-Term Preservation:** You can stash frozen foods for months or even years. It's like having a backup food plan.
- **Retains Freshness:** Some methods change how food tastes or feels, but freezing? It keeps both flavor and texture just right.

- **Convenience:** Frozen foods are super handy. Prep meals ahead of time and portion them out with ease.

Getting Started with Freezing

Ready to start? Follow these easy steps:

1. **Choose High-Quality Ingredients:** Go for fresh, top-quality stuff to get the best taste and texture when you freeze.
2. **Preparation:** Give those ingredients a wash. Get rid of peels, skins, or any unwanted bits.
3. **Packaging:** Pop your prepped items into airtight containers or freezer bags. But leave a little space at the top—things expand when they freeze.
4. **Labeling:** Keep track! Write what's inside and the date you froze it on every bag or container.
5. **Freezing:** Lay items flat in the freezer for even freezing. Single layers work best.
6. **Maintain Temperature:** Keep it at or below 0°F (-18°C). This way, your food freezes fast and stays good.

Popular Foods for Freezing

- **Fruits:** Berries, bananas, slices—all great for smoothies, treats, or baking.
- **Vegetables:** Think green beans, peas, and corn. Blanch them first to keep them fresh.
- **Meats:** Chicken breasts, ground beef, and fish fillets—perfect for soups, stews, or stir-fries.

Tips for Success

- **Blanching:** Blanch vegetables before freezing to lock in color, flavor, and nutrients.
- **Flash Freezing:** Lay out individual portions on a baking sheet before bagging them to avoid sticking together.
- **Avoid Freezer Burn:** Seal containers well! Air is the enemy—it causes freezer burn and messes with quality.
- **Thawing:** Plan ahead. Move frozen foods to the fridge overnight or use the microwave's defrost setting if you need it quick.

FERMENTATION: EXTENDING SHELF LIFE AND ENRICHING FLAVOR

Fermentation is like magic! It makes food last longer and taste better. Here, we dive into fermentation, sharing some handy tips and tricks.

Fermentation is this cool, natural process where tiny microorganisms break down carbs in foods into acids, gases, or alcohol. This not only keeps the food from spoiling but also makes it more nutritious and delicious.

Benefits of Fermentation

- **Preservation:** Fermented foods last longer because the acids produced during fermentation prevent bad bacteria from growing.
- **Flavor Enhancement:** The fermentation process adds complex flavors and aromas to foods.
- **Nutritional Boost:** Fermented foods are packed with probiotics—good bacteria that help your gut. Plus, they're full of vitamins, enzymes, and organic acids.

Getting Started with Fermentation

Ready to ferment at home? Here's how:

1. **Choose Fresh Ingredients:** Fresh is best! Use high-quality ingredients without any blemishes or bruises.
2. **Sterilize Equipment:** Make sure all jars, lids, and utensils are clean and sterilized to avoid unwanted bacteria.
3. **Prepare the Ingredients:** Wash and chop your ingredients evenly so they ferment equally.
4. **Create a Brine or Starter:** Depending on what you're making, mix water, salt, and sugar to kickstart the fermentation (check recipes for exact amounts).
5. **Packaging:** Pack your ingredients tightly in clean, airtight containers. Leave some space at the top because they expand while fermenting.
6. **Fermentation:** Seal the containers and store them somewhere cool and dark (like a pantry). Let nature do its thing!
7. **Monitor Progress:** Check on your fermenting goodies regularly. Taste them now and then to see how the flavors are developing.

8. **Storage:** Once you're happy with the taste, move them to the fridge to slow down fermentation and keep them fresh.

Popular Fermented Foods

- **Kimchi:** A spicy Korean treat made from veggies like radishes or cabbage, spiced with chili peppers, garlic, and ginger.
- **Sauerkraut:** A tangy German favorite made from fermented cabbage; loaded with vitamins C & K plus probiotics.
- **Kombucha:** This fizzy tea drink is made from sweetened tea fermented with a SCOBY (symbiotic colony of bacteria and yeast). It's got a refreshing flavor and probiotic benefits.

Tips for Success

- **Start Small:** Begin with small batches. It helps build confidence and lets you experiment with different flavors and techniques.
- **Be Patient:** Fermentation takes time. Be patient and let your foods ferment at their own pace. Check now and then to see how they're doing.
- **Experiment:** Don't be afraid to try new things! Use different ingredients, spices, and methods to craft unique fermented foods just right for you.

PICKLING: KEEPING VEGETABLES AND FRUITS FRESH IN BRINE

Pickling is a traditional way to save veggies and fruits, making them tangy and yummy for months. Here's a friendly guide on pickling with some handy tips.

What's Pickling?

Pickling means preserving foods by soaking them in brine. This mix usually contains water, vinegar, salt, and some spices. The acidic brine stops harmful bacteria from growing and gives the food a tangy kick.

Why Pickle?

- **Long Shelf Life:** Pickled veggies and fruits last way longer than fresh ones, allowing you to enjoy seasonal produce all year.
- **Versatile:** There are endless flavor combinations and ingredients you can try, making pickled foods truly customizable.
- **Nutritional Value:** Pickling changes the texture and taste a bit, but it keeps the nutrients intact, making them a healthy choice for any diet.

How to Start Pickling?

Follow these steps to pickle your fruits and vegetables at home:

1. **Choose Fresh Produce:** Pick fresh, firm veggies or fruits that have no blemishes.
2. **Make the Brine:** Mix water, vinegar, salt, and your preferred spices in a pot. Boil it, then let it cool.
3. **Prep the Produce:** Wash and trim your veggies or fruits, removing stems, seeds, or skins. Cut them into even pieces.
4. **Packaging:** Pack the produce tightly in clean jars, leaving space at the top for the brine to expand.
5. **Add Flavorings:** Add spices, herbs, garlic, or chili peppers to enhance the flavor before pouring in the brine.
6. **Pour in Brine:** Pour the cooled brine over the packed produce in jars, submerging them completely.
7. **Seal Jars:** Use sterilized lids and bands to seal the jars tightly.
8. **Fermentation:** Store sealed jars in a cool and dark place, like a pantry or cellar, for fermentation.
9. **Check Fermentation:** Look for bubbles or cloudiness as signs of fermentation.
10. **Storage:** Once satisfied with the tanginess, move jars to the fridge or cold storage.

Popular Pickles

- **Dill Pickles:** Cucumbers in a dill-flavored brine with garlic and mustard seeds; great for snacks or sandwiches.
- **Pickled Jalapeños:** Spicy jalapeños in a tangy mix; an awesome addition to tacos, nachos, or burgers.
- **Pickled Beets:** Sweet and earthy beets in spiced vinegar; perfect as side dishes or salad toppings.

Tips for Perfect Pickles

- **Experiment with Flavors:** Don't hesitate to mix different spices and herbs for unique tastes.

- **Quality Ingredients Matter:** Good vinegar, salt, and spices make all the difference.
- **Patience is Key:** Let flavors develop slowly over time for the best results.

SMOKING: ENHANCING TASTE AND PRESERVING MEATS

Smoking is a centuries-old cooking technique that gives meats a rich, smoky flavor while also keeping them fresh. In this section, we'll guide you through the basics of smoking meat and share some practical tips and techniques.

Introduction to Smoking

Smoking involves exposing meat to low temperatures and smoke from burning wood chips or charcoal. This slow, controlled method not only adds a distinct smoky flavor but also prevents harmful bacteria from growing on the meat.

Benefits of Smoking

- **Flavor Enhancement:** Smoking deepens and enriches the flavor of meats.
- **Preservation:** Low temperatures combined with smoke help preserve meats by dehydrating them and inhibiting bacterial growth, allowing longer storage without refrigeration.
- **Versatility:** Smoking works on all sorts of foods—beef, pork, poultry, fish, even veggies—so the culinary possibilities are endless.

Getting Started with Smoking

Here's how you can start smoking food at home:

- **Selecting the Right Equipment:** Whether you go for a traditional charcoal smoker, an electric smoker, or a pellet smoker, pick one that fits your needs. Consider usability in an emergency too!
- **Preparation of the Meat:** Begin with quality cuts that aren't too fatty. Trim and season them with your favorite dry rub or marinade.
- **Preparing the Smoker:** Control temperature and moisture by adding water to the smoker's pan. Place wood chips or chunks in the designated compartment; choose types that match the meat you're smoking.
- **Temperature Control:** Preheat to 200°F-250°F (93°C-121°C). Use either a built-in or external thermometer to monitor heat levels throughout the process.
- **Smoking the Meat:** Ensure good air circulation and smoke penetration by spacing out the pieces on the smoker racks. Once loaded, close the lid—let that meat smoke slow and steady! Keep adding wood chips as needed.
- **Monitoring and Adjusting:** Regularly check the temperature; adjust airflow or add more fuel as needed to keep things steady.
- **Testing for Doneness:** Use a meat thermometer to ensure it reaches safe internal temperatures.
- **Resting and Serving:** Serve the meat hot off the smoker once perfectly cooked. Remove from the smoker and let it rest for a few minutes before serving.

Popular Smoking Woods

- **Hickory:** Strong, smoky taste—great for beef, pork, and poultry.
- **Apple:** Sweet and fruity—ideal for pork, poultry, and fish.
- **Mesquite:** Bold and robust—perfect for beef and game meats.

Tips for Success

- **Patience is Key:** Smoking takes time. Let those meats slowly soak up all that smoky goodness!
- **Experiment with Wood Varieties:** Try different woods to find unique flavors. Discover your favorites!
- **Practice Food Safety:** Always follow food safety guidelines so your smoked meats are safe to eat.

ROOT CELLARING: CLASSIC COLD STORAGE FOR FRESH PRODUCE

Root cellaring's been a go-to method for centuries, perfect for keeping fruits, veggies, and other perishables in tip-top shape. By stashing your produce in a cool, dark spot, you can stretch out their shelf life and enjoy a steady flow of fresh goodies all year long. Ready to dive into the world of root cellaring? Let's get started with some handy tips and tricks!

Purpose & Perks

Root cellaring offers some pretty neat benefits:

- **Preservation:** Tuck your produce away in a root cellar, and you'll slow down the ripening process. This way, fruits and veggies stay fresh longer sans refrigeration.
- **Reduced Waste:** Got loads of seasonal produce? Store it in your root cellar to cut down on food waste and always have fresh ingredients at hand.
- **Cost Savings:** Grow your own or buy local produce and keep it in a root cellar. You'll save money on groceries, plus there's nothing like the taste of homegrown or locally sourced goodies.

Creating Your Own Root Cellar

Here's how to whip up your very own root cellar:

- **Location:** Pick a spot like a basement, underground cellar, or insulated shed. Make sure it's cool, dark & airy with temps between 32°F to 40°F (0°C to 4°C) and humidity at 85% to 95%.
- **Insulation:** Keep heat out by insulating floors, ceilings, and walls with foam board, straw bales, or packed earth.
- **Ventilation:** Pop in vents or air ducts to get air flowing and stop moisture buildup. Adding a small fan or vent pipe helps too.
- **Shelving & Storage:** Set up sturdy shelves or racks to store stuff off the floor. Use wooden crates, baskets, or burlap sacks; store everything in one layer to avoid bruising.
- **Temp & Humidity Control:** Use thermometers and hygrometers to monitor conditions. Adjust ventilation & insulation as needed for optimal storage conditions.
- **Design Extras:** Add drawers for root veggies, hooks for hanging garlic/onions & racks for drying herbs or curing meats.

Tips for Success

- **Rotate Stock:** Always use older items first to keep things fresh.
- **Check for Spoilage:** Keep an eye out for moldy/mushy spots or bad smells—remove anything that's going bad right away.
- **Keep It Clean:** Sweep/mop the floor often & disinfect shelves/containers to stave off contamination.
- **Plan Ahead:** Sync your planting/harvesting schedule with peak production times so you always have fresh produce ready.

VACUUM SEALING: AIR REMOVAL FOR PROLONGED FRESHNESS

Vacuum sealing is a relatively newer way to preserve food, at least when compared to most other methods. Here, we'll explore this amazing technique and share some practical tips.

Purpose and Benefits

Vacuum sealing offers several important advantages, including:

- **Extended Freshness:**
  Vacuum sealing creates a tight seal around food by removing air from the packaging. This protects it from moisture and oxygen, which can cause spoilage. It keeps meats, fruits, and vegetables fresh longer, allowing you to enjoy them without worrying about quality loss.

- **Prevention of Freezer Burn:**
  Vacuum-sealed packages resist freezer burn. By eliminating air, vacuum sealing preserves the texture, flavor, and nutritional value of frozen foods, ensuring they stay in top condition until you're ready to eat them.

- **Space Efficiency:**
  Vacuum-sealed packages take up less space in the fridge or freezer compared to bulky containers. Maximize your storage space and keep your kitchen better organized.

- **Prevention of Spoilage:**
  Vacuum sealing blocks microorganisms that thrive in oxygen-rich environments. This reduces food waste and saves money, allowing you to buy in bulk and store leftovers safely.

Getting Started with Vacuum Sealing

Follow these steps to begin vacuum sealing:

1. **Choose a Vacuum Sealer:**
   Get a quality vacuum sealer that fits your needs and budget. Look for features like adjustable sealing settings, a built-in bag cutter, and compatibility with different bag sizes.

2. **Select Suitable Packaging:**
   Use vacuum-seal bags or rolls designed specifically for vacuum sealers. These bags are durable, puncture-resistant, and have a textured surface for better air removal.
3. **Prepare Food for Sealing:**
   Ensure that food is dry and clean before sealing. For liquids or moist foods, pre-freeze them on a baking sheet first to prevent liquid from being sucked into the sealer.
4. **Seal the Bag:**
   Leave enough room at the top of the vacuum-sealed bag before adding food items. Trim the bag to the right size using the built-in cutter, then place the open end into the vacuum sealer.
5. **Remove Air and Seal:**
   Follow the manufacturer's instructions to remove the air and create a tight seal. Once the air is out, the sealer will heat-seal the bag automatically.
6. **Store and Label:**
   After sealing, label the bags with the contents and the date using a permanent marker. Store them in the fridge or freezer in a single layer for optimal airflow and efficient space use.

Tips for Success

- **Use Fresh Ingredients:**
  Start with fresh ingredients to ensure maximum shelf life and flavor.
- **Avoid Overfilling:**
  Leave some empty space at the top of the bag for better seals and to prevent leaks.
- **Check Seals Regularly:**
  Periodically check sealed bags for leaks or damage and reseal if needed.
- **Experiment with Different Foods:**
  Have fun! Try vacuum sealing meats, fish, fruits, vegetables—even soups and stews!

INTEGRATING METHODS FOR MAXIMUM FOOD PRESERVATION

Bringing together various techniques can be super beneficial for stretching the shelf life of your favorite foods. It lets you hit optimal results and keeps your pantry packed with a variety of delicious, nutritious options that are always ready to enjoy.

Purpose and Benefits

Combining preservation methods has some amazing perks, such as:

- **Enhanced Shelf Life:** By tapping into the strengths of different methods, you can stretch the shelf life of stored foods, ensuring they stay fresh and tasty for a longer time.
- **Diverse Food Options:** Mixing techniques allows you to preserve all kinds of foods—fruits, veggies, meats, and dairy. This approach is perfect for catering to diverse tastes and dietary needs.
- **Resource Efficiency:** Using a mix of preservation methods helps you make the most out of what you have. This can reduce waste, saving both time and money in the long run.
- **Flavor Enhancement:** Some methods, like smoking and fermentation, add unique flavors to foods, enhancing their taste appeal.

Exploring Combinations

Let's look at a few awesome combinations that might help you achieve optimal results:

- **Canning and Freezing:**
  - Start by canning seasonal fruits and veggies when they're at peak freshness. Seal them tightly in jars using the canning method to lock in nutrients and flavor.
  - For foods best enjoyed frozen—like berries and leafy greens—try freezing after blanching. This helps maintain their texture and color. Vacuum-sealing them helps too!

- **Dehydration and Vacuum Sealing:**
  - Dehydrate fruits, veggies, and herbs using either a dehydrator or an oven to remove moisture and prevent spoilage. Once dry, vacuum-seal them to keep them fresh and prevent rehydration.
  - Vacuum-seal dehydrated meats like jerky to extend their shelf life. This combo keeps your snacks tasty and nutritious for much longer.
- **Pickling and Fermentation:**
  - Start by pickling veggies and fruits in brine or vinegar. The acid helps kill harmful bacteria while adding a tangy flavor.
  - Try fermentation for probiotic-packed foods like sauerkraut and kimchi. It not only preserves but also boosts nutritional value and digestibility.
- **Root Cellaring and Vacuum Sealing:**
  - Use a root cellar (or any cool, dark place) for storing root veggies, onions, and garlic. The steady temperature slows down spoilage.
  - For even better protection from spoilage or freezer burn, vacuum-seal root veggies before storing them. This way, your produce stays fresh all year round.

Tips for Success
- **Plan Ahead:** Think about which foods are in season and time your preservation projects accordingly.
- **Label Everything:** Make sure you mark preserved foods with dates so you can keep track of freshness.
- **Experiment and Adapt:** Feel free to play around with different combinations! Adjust recipes to suit your taste buds and dietary needs.

These tips won't just help extend the shelf life—they'll enhance your culinary adventures too!

PROJECT 7: DIY EMERGENCY RATION BARS: 2400 CALORIE RECIPE

If you've ever had to run out the door and don't want to skip breakfast, you know just how useful it is to have energy bars. In the event of an emergency, these can be a great solution for rationing food and still ensuring people get the nutrients and calories they need to continue functioning.

Materials and Preparation

Before you start, you will need the following materials:

- 2 cups of oats
- 2 ½ cups of powdered milk
- 1 cup of sugar
- 3 ounces of package Jello (any flavor)
- 3 tablespoons of water
- 3 tablespoons of honey
- Air tight containers
- 1 Medium saucepan
- 1 stove/oven
- 1 mixer
- 1 large mixing bowl
- 1 roll of parchment paper
- 1 baking sheet
- 1 rolling pin

Steps

1. Preheat the oven to 200 degrees F (93 C).
2. Add the oats, powdered milk, and sugar into the mixing bowl.
3. Add the Jello, water, and honey in the saucepan, stir and boil. Once it's boiling take it off the stove and pour the mix into the bowl.
4. Mix the ingredients, if the dough becomes too dry add a tablespoon of water to make it slightly moist. Continuing adding until the dough is flakey but still has form when kneaded together.

The Initial Mixture

5. Take your parchment paper and cover the inside of the baking sheet with it, then place your mix into the pan.
6. Roll the dough down so it fills the entire pan, then cut it into squares.
7. Bake the bars for about 1½ - 2 hours.
8. Once they're baked leave the bars out to cool for about 10 minutes. Then pack them away into air tight containers.

PROJECT 8: HOW TO MAKE HARDTACK: THE ULTIMATE SURVIVAL BREAD

It may not be particularly tasting, but when food is in short supply, this is something you can make to survive until you have more options.

Materials and Preparation

Before you start, you will need the following materials:

- Large bowl
- Measuring cup
- Rolling pin
- Cutting board
- Chopsticks
- Baking sheets
- Oven mitts
- Cooling rack
- Oven
- 2 cups of All Purpose Flour
- 1½ of Salt
- ¾ cups of Water

Steps

Use the following steps to make emergency survival bread.

1. Preheat the oven to 375 degrees F (190degrees C)
2. Mix the flour, water, salt until it's a little dry. If still sticky add a fourth of a cup of flour and stir.
3. Set the dough on the cutting board and dust the surface with flour.
4. Put the dough onto the board and flatten it with the rolling pin. Roll the dough out until it's 1/3 inches to ½ inches in thickness.
5. With the dough flattened cut the dough into squares, then use the chopsticks and poke nine holes into each square, as shown.

Ready for Baking

6. Place the squares onto the baking sheet and into the oven.
7. Bake for 30 minutes, flip it over, then back it for an additional 30 minutes.
8. Remove the bread and let it cool down for about 10 minutes. Then when the bread is completely cool put into containers.

PROJECT 9: HOMEMADE PEMMICAN: THE PERFECT SURVIVAL SNACK

Pemmican provides a way for you to get your caloric intake easily and with a good bit of flavor.

Materials and Preparation

This process will take a couple of days to complete, lots of grinding and mashing up of the ingredients.

Before you start, you will need the following materials:

- ¼ a pound of red meat or fish
- ½ cup of dried cranberries
- ½ cup of rendered fat
- 1/8 teaspoon of salt
- 1/8 teaspoon of ground pepper
- 1 baking sheet
- 1 mortar and pestle
- 1 roll of plastic wrap

Steps

1. Preheat your oven to 350 degrees F (117 C).
2. Bake your meat in the oven for about 15 - 20 minutes, then turn the oven off and leave in the oven for the day to let it dry out.
3. Put the cranberries into the mortar and start crushing them up with the pestle. Grind the berries up for about 20 minutes to make sure the berries become a fine powder. Once you're done put the berries into a small bowl and set aside.
4. Take the dried meat and start the same process you did with the berries. The only difference between the two is the meat should take around 10 minutes to become powder. When done combine the meat with the berries and set to the side.

5. Take the half a cup of the rendered fat and place it into a pan. Melt the fat at a low heat and slowly stir around until it's completely melted.
6. Pour the melted fat into the bowl and stir the ingredients together. Once mixed add in the salt and pepper for a little more seasoning.
7. With all that done form the paste into strips and wrap them up in plastic wrap. For longer storage wrap it in both plastic wrap and aluminum wrap. Either way store the pemmican in a dark, dry, room temperature room.

Fresh Pemmican

PROJECT 10: LEARN HOW TO PRESERVE EGGS FOR THE LONG TERM

Eggs are one of the foods that are good for making just about anything. Making sure you have them during emergency can expand your food options.

Materials and Preparation

Before you start, you will need the following materials:

- 12 Brown Eggs
- 1 oz Hydrated Lime
- ½ gallon Glass mason jars
- 1 quart Water

Steps

1. Check the eggs for cracks or blemishes. If the egg is cracked don't preserve, if the egg is dirty then use a dry towel or cloth to wipe the dirt off. If the egg is still dirty, then don't preserve.
2. Mix one ounce of hydrated lime for every quart of water. Stir well until all the hydrated lime has dissolved.
3. Gently put the eggs into the mason jar, then add the lime solution to the jar until all eggs are under the solution.
4. Put the top on the jar, label the jar with a date, and put the jar of eggs into a fridge or cooler.

Persevered for Later

PROJECT 11: DRY MEAT AND TURN IT INTO POWDER FOR EASY STORAGE

Meat goes bad nearly as quickly as eggs. To ensure you have a reliable source of protein, you can turn some of it into jerky, and some of it into powder.

Materials and Preparation

Before you start, you will need the following materials:

- 2 pounds of pork (including bone)
- 4 Carrots (long orange)
- 6 Celery stalks
- 1 medium onion
- 1 clove of garlic
- 1 tbsp of seasoned salt
- 1 tsp of ground black pepper
- 1 blender
- 1 crockpot
- 1 baking sheet
- 1 oven
- 1 fine mesh strainer
- 1 roll of parchment paper
- 1 box of plastic bags

Steps

1. Cut fat from the pork and chop the vegetables, then put both the meat and vegetables into the crockpot.
2. Season the ingredients then fill the pot up until the vegetables and meat are submerged.
3. Boil the ingredients for about 2 hours so the pork meat is easily removed from the bone. Place the meat and vegetables into the fine mesh strainer. You can pre-heat the oven to 250 degrees F (121 C) as you wait.

In the Crockpot

4. Rinse the vegetables and meat with hot water to make sure there's no remaining fat from the pork. If there is any fat left, cut it away then put the meat and vegetables into the blender.
5. Blend the mix until it has no lumps or chunks in it. If you need to make the mix to churn better add a tablespoon of water to loosen up the mix.
6. Once you have a fine paste grab your parchment paper and your baking sheet. Place the parchment paper on the baking sheet, then spread the paste onto the paper. Make sure to keep the spread thin and even across the paper.
7. Place the paste into the oven and bake it for 2 hours. Remove it from the oven and break it up into big chunks, then put it back into the oven for another 2 hours. Repeat this process until the paste is 100% dry. Wash and dry your blender while you wait.

Dried Meat

8. Once it's dry then place the broken bits of the dried paste into the blender. Blend it until it resembles more of a powder than a cracker.
9. With your new powder be sure to store it in a plastic baggie and vacuum seal it closed.

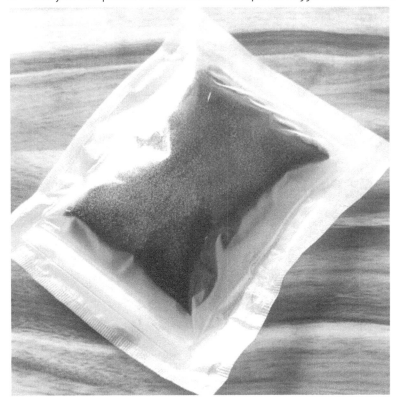

The Final Product

4. Budget-Friendly Stockpiling

RECOGNIZING THE SIGNIFICANCE OF FOOD STOCKPILING

Creating a solid stockpile of shelf-stable foods is really rewarding. Knowing your family is prepared for emergencies brings you peace of mind and a sense of independence in uncertain times. Let's dive into the details of stockpiling food and some practical ways to build a well-rounded, affordable pantry that can sustain you during a crisis.

Purpose and Benefits

Building up nutritious and shelf-stable foods has several key benefits:

- **Emergency Preparedness:** At the heart of being ready for anything is having food stored. It ensures you and your family have food when emergencies strike.
- **Self-Sufficiency:** A well-stocked pantry means less reliance on stores or disrupted supply chains. It's like having your own little grocery store at home just for yourself!
- **Cost Savings:** Buying in bulk and grabbing discounts lowers your food bills. Stockpiling can be quite economical this way.
- **Nutritional Security:** Keeping varied and balanced foods on hand means you'll always have access to important nutrients, vitamins, and minerals, keeping everyone healthy and happy.

It's reassuring to know you're prepared for anything, right? By focusing on these strategies, you can build up a fantastic food stockpile that looks after both budget and nutrition needs. So, why wait? Start building your stockpile now—you'll thank yourself later!

BUDGET-FRIENDLY TIPS FOR PREPPING

It's crucial to be both resourceful and practical, especially when stockpiling essential supplies. Being prepared doesn't have to empty your bank account. With a thoughtful approach and a little creativity, anyone can get ready without spending too much. Let's explore budget-friendly tips to build a solid plan without losing quality or effectiveness.

1. Set a Realistic Budget

First, evaluate your finances and determine how much you can comfortably spend on prepping each month. Set a budget that covers both short-term needs and long-term goals.

2. Prioritize Essential Supplies

Identify the most critical items for your plan—food, water, shelter, and medical supplies. Focus on these first before considering non-essentials.

3. Take Advantage of Sales and Discounts

Watch for sales, promotions, and clearance events at local stores, online retailers, and warehouse clubs. Buy discounted items and make bulk purchases to save more.

4. Utilize Coupons and Rebates

Clip coupons from newspapers, magazines, and online sources to save money on supplies. Look for rebate offers and cashback opportunities to stretch your dollars further.

5. Buy in Bulk

Purchase staple items like rice, beans, pasta, canned goods, and hygiene products in bulk to lower the cost per unit. Consider joining a bulk-buying group or splitting bulk buys with friends or family to get wholesale prices.

6. DIY Projects and Repurposing

Be creative with DIY projects! Repurpose items you already have into multifunctional prepping tools (e.g., water filtration, shelter construction). Explore DIY solutions for essential needs like food preservation.

7. Grow Your Own Food

Start a home garden to add fresh fruits, vegetables, and herbs to your stockpile. Growing your own food cuts costs and provides a sustainable source of nutritious produce.

8. Practice Minimalism

Adopt a minimalist mindset—focus on quality over quantity when acquiring supplies. Avoid unnecessary purchases and invest in versatile, multi-purpose items that will serve you well.

CREATING A WELL-ROUNDED FOOD SUPPLY: KEY NUTRITIONAL CATEGORIES

Having a balanced food stockpile with all the essential food groups is important. It ensures you have the necessary nutrients, vitamins, and minerals to promote both physical health and mental well-being, especially in times of need. Let's explore the key food groups to include in your stockpile for a balanced and nutritious supply.

Grains & Staples

Purpose:
Grains are the backbone of a healthy diet, offering carbohydrates for energy and nutrients like fiber, B vitamins, and minerals.

Practical Tips:
- **Stockpile:** Get a variety of grains like rice, pasta, oats, quinoa, and flour.
- **Storage:** Store in airtight containers or Mylar bags to keep them safe from moisture, pests, and spoilage.

- **Rotation:** Use a FIFO (First In, First Out) system to ensure older grains are used first and stay fresh.

Proteins

Purpose:
Proteins are vital for muscle repair, immune function, and overall health. Including various protein sources keeps you well-nourished during emergencies.

Practical Tips:
- **Sources:** Stock up on canned meats like tuna, chicken, and beef. Don't forget beans, lentils, nuts, and seeds for variety.
- **Long-Term Storage:** Vacuum-seal meats or dehydrate beans and legumes for a longer shelf life.
- **Versatility:** Pick protein sources that can be used in many recipes.

Fruits & Vegetables

Purpose:
Fruits and vegetables are packed with vitamins, minerals, and antioxidants. They boost immune function and overall health.

Practical Tips:
- **Diversity:** Mix canned, freeze-dried, and dehydrated fruits and vegetables.
- **Storage:** Keep canned items in a cool, dark spot. Store freeze-dried and dehydrated ones in airtight containers or Mylar bags.
- **Rehydration:** Soak dehydrated fruits and vegetables in water before eating or cooking.

Dairy & Alternatives

Purpose:
Dairy products offer key nutrients like calcium, vitamin D, and protein, which are good for bones and overall wellness.

Practical Tips:
- **Options:** Stock shelf-stable products such as powdered milk, powdered eggs, and shelf-stable cheese.
- **Longevity:** Choose dairy alternatives like powdered milk or plant-based milks, which last longer than fresh dairy.
- **Versatility:** Use powdered milk and eggs in cooking and baking to boost the nutrition in meals.

Fats & Oils

Purpose:
Fats and oils serve as energy sources and are crucial for hormone production and nutrient absorption.

Practical Tips:
- **Selection:** Get various fats and oils like olive oil, coconut oil, butter, and ghee for different cooking needs.
- **Storage:** Keep fats and oils in dark containers away from light and heat.
- **Considerations:** Choose shelf-stable options that last longer.

Hydration

Water:
Stock up on plenty of clean drinking water. Aim for at least one gallon per person per day—for drinking and sanitation needs.

SELECTING LONG-LASTING FOODS: KEY CRITERIA AND FACTORS

Picking the right shelf-stable foods? It's all about understanding some key points. This way, you make smart choices, ensuring your stockpile stays nutritious, yummy, and lasts for a long time.

Nutritional Content
- **Purpose:** Focus on foods with high nutritional value. This keeps you healthy and energetic during emergencies.

- **Considerations:**
  - **Essential Nutrients:** Go for foods packed with essential nutrients like vitamins, protein, and healthy fats.
  - **Whole Foods:** Pick whole grains, lean proteins, fruits, and veggies for a balanced diet.
  - **Avoid Empty Calories:** Try to avoid processed foods that have empty calories and added sugars.

Shelf Life

- **Purpose:** Opt for foods with a long shelf life. This way, you don't need to rotate or replace them often.
- **Considerations:**
  - **Longevity:** Choose foods that last long, like canned goods, freeze-dried items, and dehydrated products.
  - **Expiration Dates:** Always check expiration dates and pick those with the longest shelf life for better storage duration.
  - **Stability:** Prioritize items resistant to spoilage. Properly sealed and preserved items are the best.

Storage Requirements

- **Purpose:** Make sure the foods you pick can be easily stored without losing quality in your environment.
- **Considerations:**
  - **Temperature and Humidity:** Think about your storage area's temperature and humidity. Choose foods that can handle these conditions.
  - **Storage Containers:** Use suitable storage solutions like Mylar bags, airtight containers, and food-grade buckets to keep stuff fresh and safe.
  - **Space Efficiency:** Look for compact and stackable packaging to maximize space and keep everything organized.

Versatility and Convenience

- **Purpose:** Aim for foods that are versatile and easy to prepare. They should fit various recipes and situations.
- **Considerations:**
  - **Multi-Purpose Ingredients:** Pick ingredients that you can use in different dishes. This adds flexibility and creativity in cooking.
  - **Preparation Ease:** Go for foods needing minimal prep or cooking time.
  - **Portion Sizes:** Think about portion sizes and packaging formats that help with portion control and reduce food waste.

SMART SHOPPING: HOW TO SOURCE AFFORDABLE SUPPLIES

You know, having an affordable prepping strategy is really important for building a good food stockpile. By being smart about where and how you shop, you can get everything you need without spending too much.

Bulk Purchases

Purpose: Buy things in large amounts to save money. It's like getting more for less.

Strategies:

- **Wholesale Retailers:** Check out places like Costco, Sam's Club, or BJ's Wholesale Club. They offer lots of stuff in big packs.
- **Online Retailers:** Look up Amazon or Walmart—they deliver bulk items right to your door. Perfect!
- **Cooperative Purchasing:** Team up with family, friends, and even neighbors. Buy together to get bulk discounts and share the expenses.

Sales and Discounts

Purpose: Use sales, promotions, and discounts to make your prep budget go further.

Strategies:

- **Couponing:** Grab those coupons from newspapers, online sources, or store flyers. They're gold for saving money on prepping supplies.
- **Promotional Events:** Keep an eye out for clearance sales, seasonal discounts, and holiday specials. You can score some huge savings.
- **Membership Benefits:** Sign up for loyalty programs and rewards clubs. They send you special offers and discounts.

Local Resources

Purpose: Check out local places for cheap prepping supplies and help small businesses in your community too.

Strategies:

- **Farmers' Markets:** Go to farmers' markets for fresh veggies, canned goods, and even handmade goodies from local sellers.

- **Discount Stores:** Visit discount stores, dollar stores, or surplus outlets. They have prepping essentials on a budget.
- **Community Initiatives:** Join community swaps, bartering events, or cooperatives. You can trade or get what you need at lower costs or even for free!

OPTIMIZING STORAGE SPACE: TIPS FOR ORGANIZATION AND ROTATION

To keep your food supply neat and functional, it's important to manage your storage space. With some clever organization and rotation tricks, you can make the most of your storage area and keep everything fresh.

Use Vertical Space

Purpose: To maximize your storage by utilizing vertical space, not just horizontal.

Techniques:
- **Shelving Units:** Invest in sturdy shelves that can be adjusted to accommodate different sized containers.
- **Stackable Containers:** Choose stackable bins to neatly pile items and keep similar things together.
- **Overhead Storage:** Install racks or hooks above eye level for lightweight items like bags of rice or beans.

Categorize and Label

Purpose: Simplify inventory management by sorting and labeling everything.

Techniques:
- **Group Similar Items:** Store like items together—canned goods with canned goods, grains with grains.
- **Labeling:** Clearly label containers or shelves with contents and expiration dates to avoid confusion.
- **Color-Coding:** Consider using different colors to identify categories or indicate priority levels.

FIFO Rotation System

Purpose: Keep items fresh by using the oldest ones first (FIFO = First In, First Out).

Techniques:
- **Placement Strategy:** Place newer items behind older ones to ensure you use the older products first.
- **Regular Checks:** Frequently check your inventory to spot items nearing expiration—use them or replace them as soon as possible.
- **Rotation Schedule:** Establish a schedule to rotate items like canned or dry goods to maintain freshness.

EXTENDED FOOD PRESERVATION METHODS: MYLAR BAGS, BUCKETS, AND BEYOND

To keep your food stockpile in great shape for the long haul, you need the right storage solutions. Mylar bags, food-grade buckets, and other containers can effectively shield your supplies from moisture, oxygen, pests, and other factors that could ruin them. Let's dive into the nitty-gritty of these storage choices to make sure your prepped foods stay fresh, nutritious, and tasty.

Mylar Bags

Purpose: These bags are super popular for long-term food preservation because they're tough, have great barrier qualities, and are very flexible.

How To Use Mylar Bags:
- **Selection:** Pick food-grade Mylar bags that fit your storage needs.
- **Filling:** Put dry goods like grains, beans, pasta, or dehydrated foods inside.
- **Sealing:** Use a heat sealer to make sure no air or moisture gets in.

- **Oxygen Absorbers:** Add these to the bags before sealing to extend shelf life.
- **Labeling:** Write down the contents and date of packaging on each bag. This helps with easy identification.

Food-Grade Buckets

Purpose: These buckets are strong and secure for storing bulk food items. They protect against moisture, pests, and contaminants.

How To Use Food-Grade Buckets:
- **Cleaning:** Clean and sanitize buckets thoroughly before using them.
- **Lining:** Use Mylar bags or food-grade liners inside the buckets as an extra barrier against moisture and odors.
- **Filling:** Fill with dry goods or properly packaged foods. Make sure they're tightly sealed.
- **Sealing:** Secure lids tightly to keep an airtight environment.
- **Stacking:** Store buckets in a cool, dry place that's well-ventilated—maximize space while ensuring accessibility.

Additional Storage Options

Purpose: Consider other storage solutions like vacuum-sealed bags, glass jars, or metal containers. These can be great for different types of foods and preferences.

How To Choose The Right Storage Option:
- **Consider Food Type:** Pick containers based on what each type of food needs—such as sensitivity to moisture, aroma retention, and light exposure.
- **Evaluate Space and Resources:** Assess how much storage space you have and your budget. Consider any logistical factors when choosing storage solutions.
- **Adaptability:** Opt for versatile options that fit various types of food items and can adapt as your needs change over time.

MEAL PREPARATION AND PLANNING: CRAFTING HEALTHY DISHES FROM STORED SUPPLIES

Getting ready for emergencies means having supplies to make tasty and satisfying meals. Let's dive into some recipes and meal planning tips to enhance your culinary experience, no matter the scenario.

Understanding Recipe Adaptation

Learn how to tweak recipes to effectively use stockpiled ingredients and create fulfilling meals.

Key Points:
- **Ingredient Substitution:** Get familiar with ingredient swaps to replace fresh items with shelf-stable ones. Dried herbs can step in for fresh, and canned veggies can take the place of fresh produce.
- **Flexibility:** Embrace recipe flexibility and be open to experimenting. Use what you have on hand, tasting as you tweak flavors and quantities.
- **Recipe Modification:** Adjust recipes based on available ingredients and dietary needs. Be creative and resourceful in crafting meals that suit your preferences.

Meal Planning Strategies

Develop a meal planning strategy to make the most of your stockpile and ensure balanced nutrition in your prepping meals.

Key Points:
- **Inventory Assessment:** Regularly check your stockpile to see what's available and plan your meals accordingly.
- **Balanced Nutrition:** Plan well-rounded meals with protein, carbs, fats, vitamins, and minerals. Mix up your ingredients to ensure you get all the nutrients you need.

- **Batch Cooking:** Prepare larger quantities of meals through batch cooking to streamline meal prep. Portion out meals and freeze them for future use.
- **Menu Rotation:** Rotate meals to avoid monotony and keep your diet varied. Plan weekly or monthly menus with different recipes and flavors.

Recipe Ideas for Preppers

Check out practical and versatile recipe ideas that maximize stockpiled foods while delivering taste and nutrition.

Key Recipes:
- **One-Pot Meals:** Make hearty stews, soups, and casseroles using canned beans, veggies, and grains. Spice them up with seasonings for extra flavor.
- **Rice and Grain Bowls:** Customize rice or grain bowls with canned meats, dehydrated veggies, and shelf-stable sauces. Top them off with nuts, seeds, or dried fruits for added texture and flavor.
- **Pantry Pasta Dishes:** Quickly whip up pasta dishes using canned tomatoes, tuna or chicken, and dried herbs. Play around with different pasta shapes and sauces for variety.
- **Baking with Shelf-Stable Ingredients:** Bake bread, muffins, and cookies using flour, sugar, baking powder, and shelf-stable fats like vegetable oil or shortening. Add dried fruits or nuts for extra taste and nutrition.

KEEPING YOUR SUPPLIES READY: REVIEWING AND REFRESHING YOUR STOCKPILE

Did you know that the job's never quite done when it comes to stockpiling? Yup, you got to regularly check and update what you've got to stay ready. Let's dive into the key practices for keeping your stash fresh and reliable.

Purpose of Reviewing & Updating Your Stockpile

Taking a regular look at your stockpile isn't just busywork. Here's why it's crucial:
- **Ensuring Readiness:** Keep tabs on your stuff and restock as needed. This way, you're always geared up for emergencies.
- **Rotating Supplies:** Using the old before the new keeps your stockpile in top shape.
- **Adapting to Changing Needs:** Life happens, and things change. Update your stash to keep up.
- **Peace of Mind:** Knowing your stockpile's good-to-go lets you focus on other prep areas without worry.

Reviewing Your Stockpile

Checking in on your stockpile helps find gaps or spots that need fixing. Here's a quick rundown on how to do it right:
- **Inventory Check:** Start with listing what you've got, noting amounts and expiration dates.
- **Assess Usage:** See which items you've used and what's collecting dust. Knowing what you use helps tweak your stash wisely.
- **Inspect Quality:** Make sure nothing's damaged, spoiled, or infested by pesky critters.
- **Update Emergency Plans (just in case):** Adjust for any new threats or changes around you that might call for a tweak in your stash.

Replenishing Your Stockpile

After reviewing, it's time to top things up. Here's how to keep your supplies strong and trustworthy:
- **Purchase Essentials:** While reviewing, jot down what needs buying.
- **Follow Storage Guidelines:** Store new items properly so they stay fresh.
- **Rotate Stock:** Put new stuff at the back so you're using older items first (FIFO means "First In, First Out").
- **Consider Seasonal Needs:** Think ahead for weather changes or events needing special supplies, like extreme weather gear or holiday extras.

- **Budget Wisely:** Plan purchases smartly, juggling cost with the need to keep a solid stash.

Embracing Continuous Improvement

Staying on top of your stockpile is an ongoing task—it takes effort and dedication. By consistently reviewing and updating, you'll be ready for anything life throws at you.

5. Mastering the Great Outdoors

GETTING READY FOR THE WILDERNESS

Heading out into the wilderness? It's all about preparation. You need to know the right gear, understand your surroundings, and be mentally and physically ready.

Essential Gear and Tools for Bushcraft

Your toolkit is like your best buddy out there. Here's what you'll need:

- **A tough knife:** This will be your go-to for cutting, carving, and prepping food.
- **Firestarter:** Could be flint and steel or ferrocerium rods—got to make that fire.
- **Water purification tools:** Think boiling pots or handy portable filters.
- **Shelter materials:** Maybe a tarp or knowing how to use what's around you.

Understanding the Wilderness Environment

Get to know the local plants and animals. Know what weather to expect and watch out for any potential predators.

Mental and Physical Preparation for Wilderness Survival

You have to be tough both in body and mind. Here's how you can prepare:
- **Physical conditioning:** Go hiking, swimming, or climbing to build up endurance.
- **Mental resilience:** Practice things like mindfulness and stress management skills.

Reviewing and Updating Your Stockpile: Maintenance & Replenishment

Keep everything in top shape by regularly updating and maintaining your gear and resources.
- **Gear Inspection:** Regularly check all your stuff for wear and tear. Sharpen that knife, make sure your firestarter works, and patch up any holes in shelter materials.
- **Stockpile Replenishment:** Keep those consumables stocked. Things like water purification tablets, first aid supplies, and firestarters need replenishment. Maintain an inventory so you can restock when needed.
- **Knowledge Update:** The wilderness is always changing. Stay informed about new survival techniques, gear improvements, and changes in your preferred bushcraft locations.

SHELTER-BUILDING TECHNIQUES

Whether using what's already out there or putting something together from found materials, your shelter is your first line of defense against wild weather.

Natural Shelters: Caves and Overhangs

Understanding Natural Shelters

Natural shelters like caves and rock overhangs provide quick refuge without much work. These pre-made havens protect you from harsh elements instantly when you're short on time or supplies. But knowing how to use them wisely and safely is essential.

Finding and Assessing Natural Shelters

- **Search with Care:** Look for caves and overhangs along hillsides or rocky places. They should offer protection in strategic spots.
- **Assessment:** Once found, check if they're safe—look for animal signs like tracks or nests that might mean it's already taken.
- **Structural Stability:** Check for loose rocks or unstable parts that could be risky. Your shelter should protect, not endanger you.

Making the Most of Natural Shelters

- **Insulate Your Space:** Even caves can steal body warmth through the ground. Use dried leaves, pine needles, or clothes to make an insulating layer.
- **Ventilation & Fire Safety:** If you'll use fire for warmth or cooking, ensure good ventilation. Never light fires in closed or deep caves.
- **Securing the Entrance:** In cold places (or for extra security), block part of the entrance with branches or rocks to keep warmth in and wildlife out.
- **Leave No Trace:** Respect natural shelters by minimizing impact—leave them as they are so others (humans or wildlife) can use them too.

Considerations Before Using Natural Shelters

- **Water Source Proximity:** A nearby water source is good, but watch for flood risks in low areas.
- **Check for Hazards:** Beyond structure issues—look out for flood or landslide risks too.
- **Respect Wildlife:** If it's home to animals already, move on—don't displace them or hurt yourselves by staying put.

Debris Shelters: Making Your Wilderness Haven

As the name suggests, debris shelters are built using what's around you; they blend into surroundings, providing warmth and safety against the elements.

Choosing Your Site Carefully

- **Safety First:** Steer clear of places prone to dangers like flash floods, rockfalls, or high-wind paths.
- **Resource Accessibility:** Being near water, firewood, and edible plants improves your survival experience.
- **Ground Conditions:** Flat or slightly raised ground prevents water pooling and ensures a comfy sleep space.

Lean-to Shelter - Quick Build
- **Foundation:** Secure a long, strong branch (ridgepole) against a tree or rock—it acts as the main support.
- **Skeleton:** Place smaller branches angled along the ridgepole length to create walls.
- **Insulation + Water-Proofing:** Pile leaves, moss, or pine needles on the framework; add at least a 2-foot debris layer—the hotter or drier it gets, the more insulation you need.

A-Frame Shelter - Solid Refuge
- **Framework:** Set up two branches forming an 'A'; secure the tops, then lay the ridgepole across.
- **Ribbing:** Arrange ribs along the ridgepole in tight spaces.
- **Covering:** Layer the frame thoroughly with debris, starting from the bottom and working your way up like shingles to repel water and keep the inside dry and warm.

Crucial Elements for Debris Shelter Construction
- **Doorway Design:** Make sure your entrance faces away from the usual winds. A small entrance helps keep warmth inside the shelter.
- **Bedding Matters:** Use debris to make a bed, insulating your body from the cold ground.
- **Ventilation:** Put a small vent at the top of your shelter. This lets air flow, which is super important for managing condensation and smoke if you have a small fire for warmth.

Enhancing Your Shelter
- **Waterproofing:** Use bark or extra dense materials like mud to fill gaps and make it more water-resistant.
- **Wind Proofing:** Place heavier logs or rocks on your shelter's outer layer to shield it from strong winds.
- **Personal Touch:** Customize your shelter to fit your needs. Add a tiny storage spot for gear or create insulation pockets with more debris.

Practice and Patience
Building a good debris shelter takes practice. Understanding insulation, waterproofing, and stability principles is crucial. Start practicing in safe environments to improve your skills.

Respect for Nature
When building your shelter, remember to respect the environment. Use dead materials if possible and aim to leave little trace behind when you're done.

Step-by-Step Guide to Building a Debris Hut
A debris hut uses natural materials to create a shelter, showcasing how you can utilize your environment wisely. Here's how to do it:

Selecting Your Location
- Avoid low spots that might get waterlogged.
- Look for natural windbreaks like big rocks and thick bushes.
- Ensure you're close to resources like water and firewood.

Gathering Materials
- Get a long, sturdy branch for the spine (ridgepole).
- Find smaller branches for the ribs.
- Collect lots of dry leaves, moss, and pine needles for insulation and waterproofing.

Building the Foundation
- **Position the Ridgepole:** Secure one end on a tree fork or between two rocks, ensuring it's sturdy.
- **Ground Insulation:** Lay down a thick layer of debris on the ground for insulation.

Constructing the Frame

- **Lay the Ribs:** Position smaller branches against the ridgepole to form the hut's skeleton. Ensure they're close enough to hold debris but still allow enough interior space.
- **Cross-Bracing for Stability:** Use additional branches to create cross braces along the structure for extra stability.

Insulating the Structure

- **Start from the Bottom:** Layer leaves, moss, and pine needles over the frame, starting from the bottom up.
- **Thickness is Important:** Add at least 2 feet of debris all over to ensure good insulation and waterproofing.

Sealing the Hut

- **Entrance:** Make a small door from debris that can be pulled in after you enter to keep warmth inside.
- **Ventilation:** Leave a small gap at the highest point for air circulation.

Final Touches

- **Press Down the Outer Layer:** Walk over the debris to compact it slightly.
- **Door:** Add a door using more branches or bark pieces to seal it up nicely when you're inside.

Maintenance

- **Always Add More Debris:** Over time, insulation settles. Adding more maintains warmth and waterproofing.

Remember, how well you build your debris hut can hugely affect your wilderness experience. It can offer you a warm & dry refuge. Practice building in various places to sharpen your skills so you're ready for any situation!

Creating Long-Term Habitats

Ever wondered how to create a cozy and lasting habitat that jives perfectly with nature? It's pretty cool—a mix of hands-on skills and a dash of creativity. Let's dive into some friendly tips for building a long-term nest that's safe, comfy, and sustainable.

Selecting the Perfect Spot

- **Avoid Natural Hazards:** Skip spots prone to natural problems like floods, avalanches, or heavy winds. Check if the landscape looks stable.
- **Water Matters:** Make sure there's a clean and reliable water source close by! You'll need it for drinking, cooking, and staying clean.
- **Use Nature's Gifts:** Your chosen spot should have lots of natural goodies—think building materials, firewood, and foraging supplies.
- **Sun & Wind:** Find a place that gets enough sunlight (for warmth) and breeze (to cool). Position your habitat to harness these nifty elements.

Designing Your Habitat

- **Draw Up a Plan:** Sketch where you'll sleep, store stuff, cook, and build your fire pit.
- **Use Natural Helpers:** Take advantage of natural features like rock faces for shelter or trees for support.

Gathering Materials

- **Harvest Smartly:** Collect wood, leaves, stones—but remember to respect the ecosystem. Use deadwood and fallen branches to keep your impact light.
- **Mix It Up:** Different parts of your habitat need different materials—strong logs for frames; branches and leaves for insulation and roofing.

Building Techniques

- **Strong Base:** Begin with a sturdy foundation using big logs or stones for lasting stability.
- **Frame & Insulate:** Create a frame from big branches; fill gaps with smaller ones, leaves, and moss for insulation.
- **Roof That Lasts:** For long-term shelters, build a slanted roof with big logs covered in layers of branches, leaves, and earth to make it waterproof.
- **Use Natural Cordage:** Vines or plant-fiber cords work great to bind everything together.

Comfort and Practicality

- **Set up Fire Safely:** Plan a safe fire area with a chimney or smoke hole for venting smoke out.
- **Storage Ideas:** Build shelves from wood to keep supplies dry and tidy.
- **Cozy Sleeping Area:** Lift your sleeping space off the ground using logs layered with branches, leaves, and animal hides or blankets to stay warm.

Sustainability Touches

- **Catch Rainwater:** Create a system to gather and filter rainwater using whatever you can find or salvage.
- **Manage Waste Smartly:** Designate composting and waste areas far from living and water spots.
- **Grow Food:** If possible, start a small garden with seeds you find. It'll go well with your foraging and hunting efforts.

Keeping It in Shape

- **Inspect Regularly:** Regularly check and fix up your habitat for any wear and tear or damage.
- **Adapt As Needed:** As seasons change, tweak your habitat to keep it warm, dry, and secure. Add layers or set up windbreaks when needed.

Living With Nature

- **Leave No Trace:** Aim to minimize environmental impact. Use renewable resources; practice sustainable techniques.
- **Respect Wildlife:** Make sure your habitat doesn't bother local wildlife. Use natural deterrents if needed to keep critters at bay.

Building a long-term habitat is all about survival skills mixed with a love for nature. By planning well, living sustainably, and adding your own unique touch—you'll create a true home in the wilderness!

METHODS FOR STARTING A FIRE

Making fire is a crucial skill in the wilderness. It gives light, warmth, and lets you cook meals and clean water. Let's break down the steps and materials needed to make a fire the bushcraft way.

The Science of Fire: Understanding the Fire Triangle

Making fire requires three things: oxygen, heat, and fuel.

- **Oxygen**: It's all around us, but ensure your setup has good airflow.
- **Heat**: This comes from different methods like friction, striking a ferro rod, or focusing sunlight.
- **Fuel**: Includes tinder, kindling, and large wood pieces that keep your fire going.

Gathering and Preparing Natural Fire-Starting Materials

The first step is getting the right materials. You'll need different stages of fuel: start with tinder, then add kindling, and finally add fuel wood.

Tinder: Identifying and Processing Natural Tinder

Tinder catches fire quickly but burns fast. Here are common types:

- **Dry leaves and grasses**: Ensure they're dry. Crumble and fluff them to increase surface area.
- **Birch bark**: Contains oils that ignite easily, even when wet. Shred into fine strips.
- **Cattail fluff**: Excellent tinder; burns very quickly.
- **Dead pine needles**: Collect a bundle; they catch fire easily.

Kindling and Fuel Wood: Selection and Preparation

Once the tinder catches, you'll need kindling to build up the fire so it can burn larger wood pieces.

- **Kindling**: Look for small twigs and branches about pencil-width. They should snap cleanly, indicating they're dry. Break or cut them into small lengths. Do **not** break branches from live plants; they won't burn as well.
- **Fuel Wood**: After the kindling is on fire, add bigger pieces of wood. Ensure they're dry; dead branches off the ground work well. Chop or break them into sizes that fit your fire setup.

Building the Fire Structure

Depending on where you are and what you need the fire for, you can build different structures:

- **Tipi**: Place tinder in the center, then build a cone with kindling around it. Gradually add bigger pieces, leaving gaps for oxygen.
- **Log Cabin**: Start with a small tipi of tinder and kindling. Stack larger wood pieces around it in a crisscross pattern, like a log cabin. This structure is great for cooking and provides a stable platform.

Igniting the Fire

With your structure ready, it's time to light the tinder. Use these friction methods:

- **Bow Drill**: A classic method using a bow, spindle, fireboard, and bearing block. It takes practice but works well.
- **Ferrocerium Rod**: Strike with metal (like the back of a knife) to create sparks. Aim directly at the tinder.

Maintaining Your Fire

Add kindling and fuel wood as needed. Monitor the fire's size and heat output. Manage it carefully so it serves its purpose—whether for warmth, cooking, or signaling.

Safety Tips

- Clear the area around your fire to prevent unwanted spreading.
- Never leave your fire unattended, and fully extinguish it before leaving by using water or smothering it with dirt.
- Practice helps you make fires safely and efficiently.

Friction-Based Fire-Making Methods

Friction-based methods of making fire display human creativity and a deep connection with nature. It's all about converting kinetic energy into heat via specific material rubbing—enough to ignite tinder! Let's look at the practical steps.

Hand Drill Technique

Using just hands and some wood pieces, this method is primal and very satisfying.

Materials Needed:

- **Spindle**: A straight dry stick about 2 feet long, ideally from softwood like willow or cedar.
- **Fireboard**: A flat softwood piece with a small notch cut into its edge, plus a depression beside the notch for the spindle.
- **Tinder Bundle**: A collection of dry, fibrous material that will catch the ember from the drill.

Steps:

1. **Prepare the Fireboard**: Cut a small depression on the edge of the board and a V-shaped notch leading out from the depression.
2. **Orient the Spindle**: Place one end into the fireboard's indentation.
3. **Generate Friction**: Rapidly twirl the spindle between your palms while pressing down lightly to create friction.
4. **Create Ember**: As heat builds up, wood dust accumulates in the notch, eventually forming a small ember.
5. **Transfer Ember**: Gently place the ember in the tinder bundle and blow until it lights up.

Bow Drill Technique

The bow drill adds a bow, making the friction process less tiring and more sustained.

Materials Needed:

- Same as the hand drill, plus...
- **Bow**: A flexible, curved stick with a string (paracord or natural material).

Steps:

1. **Loop the Bowstring**: Wrap it around the spindle once.

2. **Position the Spindle/Fireboard**: Place the spindle into the depression, with the bowstring wrapped around it.
3. **Sawing Motion**: Use the bow to drive the spindle back and forth efficiently, creating an ember. Follow the same steps as the hand drill to light up the tinder bundle.

Fire Plough Method

This is a more physical technique, pushing a hard spindle through a softer board to create heat.

Materials Needed:

- **Plough Board**: A board with a groove cut lengthwise.
- **Spindle/Plough Stick**: A harder piece of wood rubbed along the groove to produce an ember.

Steps:

1. **Prepare the Plough Board**: Cut a straight groove down the center of the board.
2. **Generate Heat**: Run the spindle or stick back and forth along the groove, applying downward pressure until an ember forms.
3. **Transfer the Ember**: Place the ember carefully inside the tinder bundle and blow until it ignites.

Essential Tips for Success

- **Dry Materials Matter**: Ensuring everything is dry is crucial; otherwise, it becomes much harder to ignite a fire!
- **Practice Patience**: Practicing these techniques patiently will eventually lead to success, even if there are initial failures.
- **Quality Tinder is Key**: Fiber-rich, dry materials yield better results and ensure a successful wilderness fire.

Remember to stay safe—have some water ready—and savor the connection back to our roots as you learn valuable survival skills and experience the wilderness in all its challenging, yet magical, beauty. Each fire you make is a step deeper into nature's heart, offering lessons worth celebrating throughout our lifetimes!

Alternative Natural Fire-Starting Methods

There are a lot of cool ways to start a fire using natural materials. The most common ones are solar ignition and flint and steel sparking. Let's dive into these fun methods!

Solar Ignition Using a Magnifying Lens

Using the sun to start a fire is awesome—it saves energy and resources, too.

Materials Needed:

- **Magnifying Lens:** This could be a regular magnifying glass, binocular lenses, or the lens from a camera or smartphone.
- **Tinder:** You need some dry material that catches fire easily—char cloth, dried grass, or even a piece of paper.

Steps:

1. **Angle Your Lens:** Position the lens between the sun and your tinder. Adjust until you get a tiny beam of light focused on the material.
2. **Be Patient:** Hold steady and keep the sunlight focused on that spot. Soon enough, the heat will make your tinder smolder.
3. **Transfer Smoldering Tinder:** When you see an ember glowing, carefully move it to a larger tinder bundle. Gently blow on it to ignite.

Key Tips:

- **Optimal Conditions:** Clear and sunny days work best. The brighter the sun, the easier it will be.
- **Lens Care:** Keep your magnifying lens clean and scratch-free for the best results.

Flint and Steel: Traditional Sparking Techniques

The flint and steel method is super reliable and works in various weather conditions, making it incredibly useful.

Materials Needed:
- **Flint:** A hard rock that makes sparks—if you can't find flint, try chert, quartz, or agate.
- **Steel:** High carbon steel works great. Use an old file, knife, or special striking steel.
- **Char Cloth:** A piece of natural fabric charred in controlled burning—perfect for catching sparks.
- **Tinder Bundle:** When your char cloth catches a spark, place it into a larger bunch of dry, fibrous material to create a flame.

Steps:
1. **Prepare Your Char Cloth:** Place a small piece above the flint so it catches the sparks.
2. **Strike to Create Sparks:** Hold the steel in one hand and the flint in the other. Press the steel against the flint edge and aim the sparks at the char cloth.
3. **Catch a Spark:** When the spark hits the char cloth, watch it glow!
4. **Ignite Tinder Bundle:** Move the glowing char cloth to your tinder bundle and gently blow to get the flame going.

Key Tips:
- **Sharp Edge:** Make sure your flint has a sharp edge for striking.
- **Spark Direction:** Aim the strikes directly toward your char cloth.
- **Wind Protection:** Protect your tinder and char cloth from wind while catching sparks.

Starting fires is such an important skill with lots of uses—learning and practicing before an emergency can really make all the difference.

NATURAL WATER SOURCES

Water Sourcing & Purification

So, you're out in the wild. Finding water? It's as vital as shelter or fire. Let's dive into how to spot those natural water sources you might need.

Identifying Freshwater Sources in the Wilderness

Sometimes you've gotta be like Holmes when looking for water out there. Spotting it takes a keen eye. Here's some help:

Look for Animal Tracks

Wild creatures hang out near water spots. Follow their tracks, especially bird paths early or late in the day. They'll lead you to hydration.

Observe the Landscape

Low areas and valleys catch water naturally. And if you see lush green plants? Bingo! There's likely water nearby.

Listen for Water

In the quiet woods, a running stream or river can be heard from a distance. Just listen closely.

Key Identification Tips:
- **Clear, Flowing Water**
  Always pick moving water that's crystal clear. Stagnant pools? They're like hotels for parasites and microorganisms.
- **Avoid Contaminants**
  Steer clear of spots near factories, farms, or crowded places. Safety first!

Natural Methods for Water Purification

Now that we know where to find it, let's clean it up—wild style.

Boiling with Improvised Containers

Boiling is the champ method to zap pathogens in your water. But without a pot? No problem!

Rock Boiling

Heat some rocks in your fire (about 30 minutes). Grab them using sticks or green wood tongs and plop them into your container of water. The hot rocks will get that water boiling and safe to drink.

Bamboo Containers

Use bamboo segments as makeshift containers. Fill one with water and put it on or near the fire. Bubbles mean it's boiling and good to go!

Boiling Duration:

If you're at a low spot, boil for 1 minute; if you're high up (above 6,500 feet), boil it for 3 minutes.

Filtration Techniques Using Natural Materials

Can't boil? Let's filter instead.

Sand & Charcoal Filter

What You Need:

A bottle or hollow log, sand, charcoal from your fire (but not ash), small stones.

Construction:

1. Layer stones on the bottom for drainage.
2. Add crushed charcoal next—it soaks up chemicals and improves taste.
3. Top it off with fine sand to catch dirt.

Slowly pour water through this filter—let it drip into another container below and repeat if needed.

Grass & Moss Filter

For quicker fixes? Use grass, moss, or even fill a clean sock with these materials. Pour water through this bundle to pre-filter before any further purifying steps.

Filter Maintenance:

Make sure to clean or swap out your filters regularly—they can become little bacteria farms otherwise!

Final Steps for Safety:

Even after filtering, purifying with methods like chemicals or UV light hits those tiny microbes that filters might miss.

Mastering how to find and clean water without store-bought tools makes you feel more at home in nature! Skills like these boost your resilience and independence outdoors. Remember—water is life! Handle it with respect and care—it'll sustain you through all your wild adventures.

6. Essential Skills for Outdoor Exploration

Navigation is a wilderness skill that can be, literally, lifesaving. You'll need to move away from your shelter and find your way back, meaning you need to get oriented without modern luxuries. Master this skill, and you gain the knowledge to explore vast regions not often traveled by others. This chapter covers how to use natural navigation—using landmarks, the sun, and stars to guide you.

THE ESSENCE OF NATURAL NAVIGATION

The wilderness doesn't come with GPS or street signs. Instead, it offers a rich tapestry of natural markers and celestial bodies as guides for those who know how to read them.

Using Landmarks & Topography

Landmarks and topographical features are Earth's navigational aids. They offer reference points in vast, open spaces.

- **Identify Prominent Landmarks:** Mountains, rivers, uniquely shaped trees—these can serve as waypoints. Keeping these in sight helps maintain your bearings.
- **Read the Terrain:** Valleys lead to water, ridges give vantage points, and slopes can indicate cardinal directions based on the sun's path.

Key Techniques:
- **Map Reconnaissance:** Before venturing out, study maps of the area to familiarize yourself with key landmarks. Always carry hard copies—they're far more reliable than battery-dependent tech.
- **Visual Bearings:** Pick a landmark in your intended direction of travel and head towards it. Choose new landmarks as you progress.

Understanding Sun Position and the Shadow Stick Method

The sun is a reliable compass if you know how to read it correctly.

- **Sun Position:** In the northern hemisphere, the sun travels from east to west while arcing southward. Use this to approximate direction during the day.
- **Shadow Stick Method:** Insert a stick vertically into the ground. Mark the tip of its shadow. Wait 15-30 minutes and mark the new shadow tip location. Draw a line between these marks for an east-west indicator; the first mark signifies west.

Practical Application:

Use the shadow stick method for a makeshift compass when lacking mechanical devices.

Night Navigation Using Stars

The night sky brims with stars and constellations that guide nighttime travelers.

- **The North Star (Polaris):** In the northern hemisphere, finding Polaris aligns you with true north. Spot the Big Dipper; follow its outer edge line until you reach Polaris—the bright star at this line's end.
- **The Southern Cross (Crux):** In the southern hemisphere, extend the long axis of Crux four times to locate south.

Stellar Navigation Tips:

- **Familiarize Yourself with Constellations:** Knowing key constellations aids navigation while making stargazing more pleasant.
- **Practice at Home:** Use star maps and apps to learn about night skies before relying on them in actual field conditions.

Enhancing your wilderness navigation skills is about preparation, observation, and adaptation. It's more than just reaching a destination; it's about confidently traversing the wild.

ENHANCED NAVIGATION STRATEGIES

When you're out in the wild without modern or traditional tools, advanced orientation techniques can be a lifesaver. Let's dive into how you can navigate when you don't have the tech humans have been using for ages.

Reading Natural Signs & Animal Behavior for Direction

Mother Nature offers plenty of clues to help guide your way.

- **Observing Growth:** Trees in the northern hemisphere usually grow denser on their northern side to soak up more sunlight from the south. This might remind you of a compass.
- **Watching Animal Movements:** Many animals have regular paths leading to water or shelter. Observing these tracks could guide you toward essential resources and help with your orientation.
- **Utilizing Moss Growth:** Moss often grows on the damp, shaded sides of trees and rocks. In many places, that's typically the northern side. This gives you a rough sense of direction.

How to Apply:

- **Practice Observational Skills:** Get into the habit of watching natural signs in your local area. This sharpens your ability to read them when you're in unknown terrains.
- **Cross-Reference Signs:** Use multiple indicators from nature to double-check your direction. This ensures more accuracy in your orientation.

Constructing Improvised Compasses

Out in the wilderness, you can use creativity and simple materials to find your way.

Water Compass

Materials Needed: Some water, a leaf, and a small bit of metal (like a needle).

Procedure:
1. **Magnetize the Needle:** Rub the needle with silk or wool to magnetize it.
2. **Place on Leaf:** Put the needle on the leaf.
3. **Float on Water:** Gently place the leaf in still water. The leaf will rotate until the magnetized needle points north-south.

Shadow Compass

Materials Needed: A stick and some sunshine.

Procedure:
1. **Stick Placement:** Push a stick vertically into the ground.
2. **Mark Shadow Tip:** Mark where the shadow tip ends.
3. **Wait and Mark Again:** After 15 to 30 minutes, mark the new shadow tip position.
4. **Draw a Line:** Connect both marks to form an east-west line. Stand with the first mark (west) on your left, and you'll be facing north.

Key Techniques:
- **Ensure Accuracy:** Test and calibrate your homemade compasses where you know directions beforehand to ensure they're precise.
- **Use with Natural Signs:** Combine your makeshift compass with natural signs for more accurate navigation.

Remember, in the wilderness, you're never lost if you know how to find your way.

Understanding Maps & Compasses

Maps are vital tools for navigation. They show detailed representations of terrains, landmarks, and features essential for planning routes and getting oriented. Topographic maps are especially useful, offering information about elevation, contours, water bodies, and vegetation. Knowing how to interpret map scales, legends, and symbols is crucial for understanding maps accurately.

Compasses are reliable navigational aids that help determine direction, take bearings, and maintain course orientation. A basic compass has a magnetic needle aligned with Earth's magnetic field pointing northward. By aligning this needle with map grids or landmarks, you can establish your position and navigate effectively.

Map Reading Techniques

Mastering map reading is essential for effective navigation. Learn these skills to understand your surroundings and plan routes efficiently:

- **Orientation:** Line up your map with the surrounding environment so features match up correctly.
- **Grid References:** Use grid lines and coordinates to pinpoint specific locations precisely on your map.
- **Scale Interpretation:** Understand map scales to estimate distances accurately and gauge feature sizes properly.
- **Contour Understanding:** Read contour lines to visualize changes in elevation like slopes, valleys, and ridges.
- **Feature Identification:** Recognize both natural and artificial features shown on maps—things like rivers, roads, and buildings.

ESSENTIAL NAVIGATION TIPS

You know, besides just knowing how to read and use a compass, you can pick up some practical tips to make navigation faster and safer. Here are a few to keep in mind:

- **Maintaining Situational Awareness:** Always keep your eyes peeled for what's around you—landmarks, terrain features, and hazards. This helps you navigate and stay safe.

- **Using Handrails and Catching Features:** Look for noticeable linear features like rivers or trails to guide you. Also, big features like hills or buildings can confirm you're on the right path.
- **Taking Bearings and Back Bearings:** Use your compass to find direction towards particular features on the map. Back bearings are handy too—just look back at known points from where you are to double-check your position.

Navigation Strategies & Route Planning

Good navigation involves planning smartly and thinking about different factors like terrain, distance, aids, and safety. Here are some strategies:

- **Choosing Safe and Efficient Routes:** Check out terrain conditions and any obstacles when picking routes. Go for paths that balance safety and speed toward your goal.
- **Utilizing Handheld GPS Devices:** While knowing map and compass basics is key, handheld GPS devices can be super helpful, especially when things get tricky or exact locations matter.
- **Creating Waypoints and Checkpoints:** Set up waypoints and checkpoints along your route. They help track progress and keep you on course. Use standout features or GPS coordinates as reference points.

7. Emergency Medical Skills for Survival

ESSENTIAL FIRST AID TECHNIQUES

Treating Cuts

Cuts are super common, and it's super useful to know how to take care of them properly. This helps keep infections away and avoids more problems.

Procedure:

1. **Clean the Wound:** First, wash the cut with soap and water to get rid of dirt or anything nasty.
2. **Apply Pressure:** Stop the bleeding by pressing down on the cut with clean gauze or a cloth.
3. **Bandage the Wound:** Once the bleeding stops, cover it with a sterile bandage to keep germs out.
4. **Monitor for Infection:** Keep an eye on the wound. If you see redness, swelling, or pus, it's time to see a doctor.

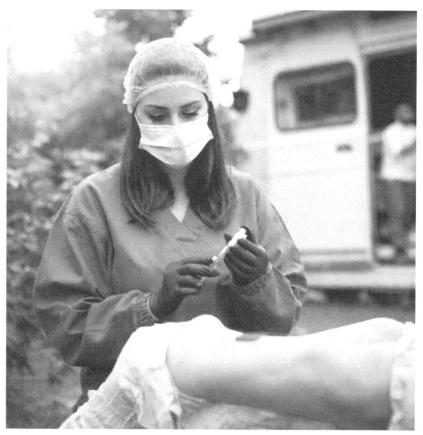

Treating Burns

Burns can be minor or pretty bad, so treating them quickly is key to avoiding more skin damage.

Procedure:

1. **Cool the Burn:** Run cool water over the burned area for at least 10 minutes. This helps ease pain and reduce swelling.
2. **Remove Clothing:** Gently take off any clothes or jewelry around the burn to avoid further harm.
3. **Apply a Cool Compress:** For minor burns, use a cool damp cloth or sterile gauze to soothe the skin.
4. **Cover the Burn:** After cooling it down, cover the burn with a sterile bandage or nonstick gauze to keep it safe from infection.

Treating Sprains

Sprains hurt! They happen when ligaments stretch too much or tear. Treating them right can make you feel better faster and heal quicker.

Procedure:

1. **Rest the Injured Joint:** If needed, use a brace or splint to keep the joint still and avoid putting weight on it.
2. **Apply Ice:** Use an ice pack for 15-20 minutes every couple of hours to reduce swelling and pain.
3. **Compression:** Wrap an elastic bandage around the joint. This helps support it and reduces swelling.

4. **Elevate:** Keep the injured limb raised above your heart. This helps reduce swelling too.

Strategies for Assessing Injuries

Knowing quickly how serious an injury is can be lifesaving. Here's how to get better at assessing injuries.

Primary Survey:
- Check Airway, Breathing, & Circulation (the ABCs) fast.
- Look for problems like choking or trouble breathing.
- Assess breathing: rate, depth, effort – any distress?
- Check pulse and overall condition – skin color, temperature – what's their circulation like?

Secondary Survey:
- After the ABCs, examine other injuries closely.
- Check for any deformities, swelling, or discoloration.
- Gently press around to detect tenderness, instability, or a grating feeling (could indicate a fracture).
- Ask about their pain level and other symptoms. More information helps with assessment.

PUTTING TOGETHER A COMPLETE FIRST AID KIT

Including Essential Medical Supplies

So, you're thinking about putting together a solid first aid kit, huh? Nice! It's super important to make sure you've got all the essential medical supplies. These tools can handle all sorts of injuries and emergencies. Let's go over what you need:

Bandages:
- **Adhesive Bandages:** Usually called "band-aids," these are must-haves for small wounds, blisters, and abrasions. They speed up healing and keep dirt and germs out.
- **Sterile Gauze Pads:** Great for bigger wounds or as a first layer under an adhesive bandage. They soak up blood and other fluids while keeping the area clean.
- **Elastic Bandages:** Also known as "ACE bandages." Useful for sprains, strains, and joint injuries. They reduce swelling and help support injured areas.

Antiseptics:
- **Antiseptic Wipes:** These wipes, soaked in alcohol or benzalkonium, are good for cleaning the skin around a wound to lower infection risks.
- **Iodine Swabs:** Single-use swabs with povidone-iodine solution. They're strong antiseptics, perfect for larger wounds or prepping skin before minor procedures.
- **Antibacterial Ointments:** Containing ingredients like bacitracin or neomycin. These stop infections and help minor cuts and burns heal quicker.

Splints:
- **Foam or Inflatable Splints:** Lightweight and easy to carry. They keep injured limbs stable (like fractures) and reduce pain until you get medical help.

Medications:
- **Pain Relievers:** Over-the-counter meds like acetaminophen (Tylenol) or ibuprofen (Advil). Useful for minor pains or fevers.
- **Antihistamines:** Medications like diphenhydramine (Benadryl) ease allergic reactions and itching.
- **Aspirin:** Good for pain and fever, but also handy during heart attacks as it thins the blood and helps prevent clots.

Tools & Instruments:
- **Scissors:** Sharp, stainless steel scissors for cutting bandages, dressings, or even clothing if needed.
- **Tweezers:** Fine-tipped tweezers to remove things like ticks or splinters stuck in the skin.

- **Thermometer:** A digital thermometer provides accurate temperatures to spot fevers or hypothermia.
- **Tourniquet:** In cases of severe bleeding, a tourniquet stops blood flow temporarily to an injured limb.

KEEPING THE KIT FULLY STOCKED AND READILY ACCESSIBLE

Once you've collected all the important supplies, it's critical to keep your first aid kit well-organized and easy to reach when you need it most.

Organization:

Arrange your medical supplies in a logical order within the kit. Group similar items together and label compartments for quick identification—this can really help. Use clear plastic pouches or zip-lock bags to separate and protect individual items, keeping everything intact and right at your fingertips.

Accessibility:

Store your first aid kit in a designated spot that's easy to get to for everyone in your home or group. Find a place that's central, visible, and easy to reach during emergencies. Also, consider having extra small kits or portable pouches for when you're on the move. Keep these mini-kits in backpacks, cars, or travel bags. This way, you're always ready to handle medical needs, no matter where you are.

Regular Maintenance:

Make sure to check and restock your first aid kit regularly. This ensures all supplies are up-to-date and in good shape. Tools with fine tips are great for removing ticks, splinters, and other pesky items stuck in the skin. Do regular inventory checks to see if everything is there and working well. If something's broken or missing, replace it right away to keep the kit fully functional.

INSTRUCTIONS FOR CONDUCTING CPR & ESSENTIAL LIFE SUPPORT

Knowing When and How to Administer Chest Compressions and Rescue Breaths

Performing Cardiopulmonary Resuscitation (better known as CPR) and basic life support means being the HERO in a cardiac emergency. These vital skills help individuals jump in promptly and effectively when someone faces sudden cardiac arrest, ensuring oxygen-rich blood keeps circulating until advanced medical help arrives.

1. Assessing the Situation:

- **Examine the Scene:** Make sure the area is safe for you and the patient before rushing in. Watch out for any dangers—like electrical, fire, or traffic hazards.
- **Assess Reactivity:** Give a gentle shake and ask, "Are you okay?" If there's no response, call emergency services (911 or your local number) right away.

2. Checking for Breathing:

- **Look, Listen & Feel:** Bring your head close to theirs and tilt it back slightly. For up to ten seconds, feel for breath on your cheek, listen for any sound, and watch their chest rise or fall.
- **Begin Rescue Breaths:** If they're not breathing normally, seal their mouth with yours, pinch their nose shut, and give two quick breaths (about a second each). Watch the chest lift as you breathe.

3. Initiating Compressions:

- **Position Your Hands:** Kneel by their side; place the heel of one hand in the center of their chest, between the nipples. Interlock the fingers of your other hand on top.
- **Perform Compressions:** Keep your arms straight—with shoulders right above your hands—then push rapidly (100-120 compressions per minute). Allow full recoil between compressions.

4. Alternating Between Compressions & Breaths:

- **Follow the 30:2 Ratio:** After 30 compressions, give two breaths. Keep this up until help arrives or they show signs of life like breathing or movement.

Understanding the Importance of Early Intervention in Cardiac Emergencies

In cardiac emergencies, time is GOLDEN. Starting CPR and basic life support immediately can save lives by keeping blood flow and oxygenation going to the brain and vital organs until advanced medical care arrives.

Minimizing Brain Damage:

- **Maintaining Oxygen Supply:** CPR helps keep blood flowing artificially—delivering oxygen to the brain and preventing permanent damage.
- **Buying Time:** Every minute without CPR drops survival chances by 7-10%. Quick action buys precious time until defibrillation or advanced help kicks in.

Increasing Survival Rates:

- **Improving Outcomes:** Studies show bystander CPR can double or even triple survival chances from sudden cardiac arrest. Acting fast boosts the odds of positive results BIG time.
- **Empowering Communities:** Training more people equips communities with crucial skills—enabling effective responses during cardiac emergencies and saving many lives!

GUIDELINES FOR HANDLING MEDICAL CRISES IN ISOLATED SETTINGS

Tips for Managing Medical Emergencies in Remote Environments

Navigating medical emergencies out in the wild can be a bit scary, but with the right mindset and some handy strategies, you can take control and provide the care that's needed. Here are some tips to help you manage these situations effectively:

Making the Most of What You've Got

First Step: Assess

Quickly and calmly assess the situation. List all available resources—like any medical gear, supplies, and people around who can help.

Set Priorities

Decide what needs attention first based on urgency and available resources. Tackle life-threatening issues right away—such as heavy bleeding, breathing difficulties, or unconsciousness.

Improvise When Needed

Sometimes, you'll have to get creative with what you have. No commercial splint? Use a strong stick or even a rolled-up magazine to stabilize a fractured limb.

Maximize Efficiency

Use everything wisely. Don't waste supplies—consider reusable items and alternatives when appropriate.

What to Do First

Look at the Patient

Start with the ABCs: Airway, Breathing, Circulation. Is their airway clear? How's their breathing and circulation? Take a close look.

Take Immediate Action

Address any immediate life threats. Stop severe bleeding with direct pressure or a tourniquet, open the airway if needed, and be prepared to perform rescue breaths or CPR.

Stabilize Them

Once you've handled the major dangers, stabilize the patient. Splint fractures and secure injured limbs to prevent further injury. Try to position them comfortably as well.

Monitor Vital Signs

Keep an eye on their temperature, blood pressure, breathing rate, and pulse. Document any changes and adjust your treatment as necessary.

Stay in Touch

Communicate Clearly

In remote environments, clear communication is crucial for getting additional help quickly. Use radios, satellite phones, or any available means to stay in contact with emergency services or other responders.

8. Emergency Sanitation

GUIDE TO PORTABLE SANITATION SOLUTIONS

Ensuring proper sanitation is just as important as securing food and water. When traditional plumbing systems fail, portable sanitation options become essential. Here's a look at portable toilets, waste bags, and composting toilets—their pros, cons, and where they work best.

Portable Toilets

Portable toilets are handy and hygienic for waste disposal in outdoor or off-grid settings.

Pros:

- **Portability:** These are lightweight and easy to move. You can set them up anywhere you like.
- **Privacy:** Many come with privacy screens or enclosures, giving users a bit more discretion.
- **Ease of Use:** They often snap together easily and use disposable waste bags, making cleanup hassle-free.

Cons:

- **Limited Capacity:** They have a limited capacity and may need frequent emptying, especially in busy areas.
- **Odor Management:** Without good ventilation or odor control additives, they can get smelly, especially in warm weather.
- **Maintenance:** Regular cleaning and disinfection are needed to keep things sanitary and bacteria-free.

Waste Bags

Waste bags are versatile and compact—perfect for emergency situations.

Pros:

- **Compactness:** Lightweight and small, they're ideal for bug-out bags, emergency kits, or camping gear.
- **Disposable:** After use, you can seal them securely and throw them into designated waste bins, minimizing environmental impact.
- **Versatility:** They can also work as emergency toilet liners or makeshift trash bags.

Cons:

- **Odor Management:** Some have odor-neutralizing additives, but others might smell bad when full or exposed to heat.

- **Durability:** If they're thin or low quality, they can tear or leak under pressure—so handle with care.
- **Storage:** Keeping large amounts of waste bags could be tricky, particularly in small spaces.
-

Composting Toilets

Composting toilets, oh these wonders! They provide a sustainable and earth-friendly alternative to your typical flush toilets. By using natural processes, they break down human waste into compost.

Pros:

- **Sustainability:** These magical toilets reduce water use and cut down pollution. They turn waste into nutrient-rich compost. That's amazing!
- **Self-contained:** Many of these models are self-contained, needing only minimal infrastructure. Perfect for remote or off-grid spots.
- **Odor Control:** When managed properly, composting toilets control odors superbly using aerobic decomposition techniques. Keeps the indoor air fresh!

Cons:

- **Initial Cost:** Yes, at first, they might be a bit pricey to buy and set up compared to conventional toilets. This can be a concern for those watching their budget.
- **Maintenance:** Regular upkeep is needed. Turning the compost now and then, and occasionally emptying it ensures it works great and stops odors.
- **Space Requirements:** They usually need some space for installing, storing compost, and ventilation. This might be tricky in tight spots.

GUIDELINES FOR EFFECTIVE WASTE MANAGEMENT

Containment, Treatment, and Safe Disposal of Human Waste

Proper sewage management is super important for keeping things clean, stopping the spread of diseases, and protecting our planet. Whether you're in a tough spot or just looking to be more self-sufficient, having good waste management practices is a must. Here's some advice to help you manage human waste better:

1. **Containment:**
    - **Designated Waste Area:** Pick a spot for waste. Keep it away from water sources and where people live. It should be close enough to access but hidden enough for privacy.
    - **Digging Latrines:** Make a hole about 6 to 8 inches deep. Cover it with something sturdy like a lid or seat for comfort and hygiene.
    - **Portable Toilet Solutions:** Consider using portable or composting toilets. These are great for long-term use and are designed to keep waste contained while reducing odor and environmental impact.
2. **Treatment:**
    - **Composting:** If possible, composting human waste is an eco-friendly option. Use a compost bin or chamber and mix in materials like sawdust or straw to aid in decomposition.
    - **Chemical Treatments:** You can use chemicals to reduce odor and break down waste in portable toilets or tanks. Follow the manufacturer's instructions for best results.
    - **Natural Decomposition:** Outdoors, nature's way can work over time. Just ensure that waste decomposes without contaminating water sources or causing health issues.
3. **Safe Disposal:**
    - **Burying Waste:** After use, bury or cover the waste promptly to keep flies, pests, and contamination away. Use soil or biodegradable materials to speed up decomposition and mask odors.
    - **Avoiding Water Contamination:** Protecting water sources is crucial! Ensure waste areas are uphill and far from wells, rivers, and other water bodies.

- **Regular Maintenance:** Keep your systems functioning properly with regular checks and maintenance. Empty toilets and compost bins as needed, and replenish supplies like sawdust or chemicals.

Minimizing Environmental Impact and Health Risks

Good sewage management keeps people healthy and the environment clean and unpolluted! By following these tips, you'll minimize any negative effects on nature and health risks:

- **Promote Decomposition:** Support natural processes by adding organic materials and allowing sufficient time for waste to break down.
- **Dispose of Chemicals Properly:** If you're using chemicals, ensure they're disposed of correctly to avoid contaminating soil and water.
- **Practice Responsible Waste Disposal:** Place solid waste in appropriate locations according to local regulations.
- **Educate and Raise Awareness:** Share the importance of proper sewage management with others and encourage responsible waste disposal.

Remember, being smart about handling waste helps keep everyone healthy, sanitary, and sustainable no matter where you are!

CRITICAL ROLE OF MAINTAINING PERSONAL HYGIENE

There are a bunch of important actions you need to keep doing to ensure good hygiene. Let's dive into why hygiene matters and some practical tips for staying clean, even in tough situations.

Emphasizing Handwashing, Bathing, and Proper Waste Disposal

- **Handwashing:** It's simple yet super effective! It's like the hero of personal hygiene. Regularly wash your hands with soap and water for at least 20 seconds. Do this before you handle food, after restroom visits, and when you cough or sneeze. Proper handwashing kicks out germs and stops the spread of infectious diseases.
- **Bathing:** Sometimes, access to showers or baths might be hard during emergencies. But guess what? You can still stay clean. Take a sponge bath using a basin of warm water and soap to clean your body. Focus on spots like the groin, feet, and armpits that sweat a lot and gather bacteria.
- **Proper Waste Disposal:** Dispose of waste responsibly to stop contamination and illness from spreading. Use designated waste areas and follow local waste management rules. If proper facilities aren't available, bury human waste (including feminine products, bandages, etc., with bodily fluids) in a shallow hole at least 200 feet away from water sources, campsites, and food prep areas.

Educating on the Role of Personal Hygiene in Maintaining Overall Health

Personal hygiene isn't just about being clean; it's about keeping yourself and others safe from illness and disease. By practicing good hygiene habits, you can reduce the risk of infections, gastrointestinal diseases, and skin conditions.

Preventing illness through proper hygiene practices helps conserve medical resources and preserve your health in tricky times. So stay positive! By taking these small steps every day, you're making a big difference in your health and well-being.

APPROACHES TO SUSTAINING HYGIENE IN DIVERSE SETTINGS

Temporary Shelters, Makeshift Camps, & Off-Grid Living Situations

When staying in temporary shelters or makeshift camps, things can get messy fast without proper care. Set up specific spots for waste disposal, handwashing, and bathing. This helps keep the place clean and reduces contamination.

Clean and disinfect communal areas regularly. Wipe or spray frequently touched surfaces like doorknobs, tables, and bathroom fixtures. This simple act can stop germs from spreading.

Practical Solutions for Keeping Clean and Staying Healthy

Hand Sanitizer: Always have alcohol-based hand sanitizer within reach. When you can't access handwashing facilities, it's your best friend. Just rub it on your hands thoroughly until dry to zap those germs away.

Hygiene Kits: Make sure every family member has a hygiene kit with essentials like soap, toothpaste, toothbrushes, and menstrual hygiene products. Keep these kits in waterproof containers or sealable bags to ensure they stay clean and dry.

By focusing on handwashing, bathing, and smart waste disposal practices, you can avoid getting sick and promote overall health. These might sound like little things, but implementing strategies such as setting up sanitation areas, using hand sanitizer, and having hygiene kits ready can help keep YOU and your loved ones healthy even in tough times.

9. Outdoor Cooking

OUTDOOR COOKING?

Yep, it's something you'll probably enjoy. Maybe you already have an idea of what to do. Lots of folks just love it. But hey, cooking in a crisis is sort of different from grilling on a sunny weekend! Let's dive into the various ways to cook food out in the wild.

Why Bushcraft Cooking Matters for Preppers

Survival Preparedness:

Knowing how to cook outdoors ensures you can feed yourself and your loved ones no matter what happens.

Resourcefulness:

Bushcraft cooking teaches you to adapt and use natural resources effectively. From foraging to making your own cooking gear, you'll learn how to thrive in any setting.

Connection to Nature:

When you cook outdoors, you're immersed in the sights, sounds, and smells of nature.

Benefits of Open-Fire Methods and Solar Cooking Techniques

Open-Fire Cooking

- **Versatility:**
  Whether you're roasting, boiling, or grilling, open-fire cooking offers endless possibilities! With a good fire, you can make all sorts of dishes using hardly any equipment.

- **Flavor Enhancement:**
  There's just something extra special about food cooked over an open flame. The smoky aroma and charred edges give your meals depth and richness.

- **Traditional Wisdom:**
  You'll pick up valuable skills handed down through generations. These skills will enrich your bushcraft knowledge.

Solar Cooking

- **Sustainability:**
  Solar cooking helps reduce reliance on fossil fuels and shrinks your carbon footprint.

- **Efficiency:**
  Solar ovens are shockingly efficient! They get hot enough to cook food pretty well. With good insulation and the right positioning, you can make tasty meals with just sunlight and some patience.

- **Low Maintenance:**
  Unlike regular cooking methods, solar cooking requires hardly any supervision. Set up your solar oven correctly and let Mother Nature do her thing while you tackle other tasks.

TECHNIQUES FOR COOKING WITHOUT CONVENTIONAL KITCHEN APPLIANCES

Overview of Open-Fire Cooking Methods

Open-fire cooking methods not only provide sustenance but also add a dash of adventure and pure satisfaction to your time in the wild. Let's explore the various methods:

Roasting: Roasting is one of the oldest, simplest ways to cook over an open flame. Skewer food—be it meat, veggies, or fruits—and cook it right over the fire. The heat from the flames chars the outer layer, giving a delicious smoky flavor while keeping it juicy inside. Endless experimenting with seasoning and marinades? Absolutely!

Boiling: Boiling is basic: immerse food in boiling liquid (usually water) to cook it. Ideal for soups, stews, pasta, and grains when you're outdoors. Use a pot suspended over the fire to control temperature and cooking time easily. This method softens tough ingredients and extracts flavors for tasty meals.

Steaming: For a gentle touch, steaming preserves natural flavors and nutrients while keeping food moist and tender. Create a makeshift steamer by placing a heatproof dish or rack over boiling water in a pot. As water evaporates, steam cooks the food evenly. Great for delicate ingredients like fish, veggies, and dumplings.

Braising: Combine searing and slow-cooking with braising to create rich dishes full of flavor. Start by searing meat or vegetables in a hot pan to caramelize surfaces. Then, simmer in broth, wine, or sauce until tender and infused with robust flavors. Perfect for tougher cuts of meat and root vegetables.

Grilling: Grilling means cooking food right over an open flame or hot coals. Place your food on a grill grate (or fire) and flip occasionally for even cooking. Easy and suitable for most foods!

Pit Cooking: This traditional method involves digging a hole in the ground lined with hot stones. Place food on top; cover with more stones, soil, and insulating materials like leaves or wet burlap. Slow-cook for tender and flavorful results—especially great for meats and root vegetables.

Campfire Dutch Oven Cooking: A Dutch oven is a heavy cast-iron pot with a tight-fitting lid—perfect for slow cooking, simmering, and baking. Distribute heat evenly by placing hot coals beneath and on top of the oven. Prepare anything—from stews to casseroles to bread and desserts.

Smoking: Smoking infuses food with flavor by exposing it to wood chip smoke. Ideal for meats, fish, and cheeses, providing that iconic smoky taste and tenderness. Use either a smoker or create one by placing food on a grill rack above fire-infused wood chips.

Campfire Skillet Cooking: Quick and convenient! Heat a cast-iron skillet over the fire, add meats, vegetables, or eggs, and cook away! The skillet distributes heat evenly, so you can sear, sauté, or fry to perfection. From bacon and eggs at breakfast to hearty stir-fries for dinner—it's incredibly versatile.

Each open-fire method brings unique perks allowing culinary creativity to flourish in nature's embrace!

Step-by-Step Guides for Preparing Meals

Cooking in the great outdoors can be a delightful adventure! A clear plan and the right tools make it enjoyable. Let's walk through an easy step-by-step process to prepare meals with improvised outdoor cooking equipment.

1. Gather Ingredients:

Start by collecting all the ingredients you need. Keep it simple and focus on fresh, high-quality ingredients that are easy to transport and cook.

2. Prepare Your Cooking Area:

Pick a good spot for your cooking fire. Clear any debris away and make a flat, stable surface using rocks or soil. If you're using a fire pit, make sure it's well-constructed and surrounded by a circle of stones to keep the flames contained.

3. Build Your Fire:

Gather dry tinder, kindling, and larger fuel wood. Place the tinder in the center of your fire pit and arrange a teepee or log cabin structure around it with the kindling and fuel wood. Light the tinder, and gradually add more fuel wood as the fire grows.

4. Set Up Your Cooking Equipment:

While your fire is burning, set up your cooking equipment nearby. If you're using a tripod, set it up over the fire pit and hang your pot or pan from the crossbar using paracord or metal wire. If you're roasting food on skewers, sharpen the ends and stick them in the ground around the fire.

5. Prepare Your Ingredients:

Wash and chop your ingredients. Remember that outdoor cooking might require larger cuts and longer cooking times. Season your food with salt, pepper, herbs, and spices to enhance the flavors.

6. Start Cooking:

Once your fire has turned into hot coals and the flames have reduced, begin cooking. Place your pots, pans, or skewers over the fire, adjusting their height to control the temperature. Monitor your food closely, rotating occasionally for even cooking.

7. Monitor and Adjust:

Keep an eye on your food as it cooks. Adjust the position of your cooking vessels and add more fuel wood if needed to maintain steady heat. Use a fireproof glove or tongs to handle hot pots and pans safely.

8. Test for Doneness:

Use a meat thermometer or the touch test to check if your food is done. Ensure meat is cooked to a safe internal temperature and vegetables are tender. Outdoor cooking might require longer times than indoor methods, so be patient!

9. Serve and Enjoy:

Once cooked, remove your meal from the fire and transfer it to serving dishes. Now, enjoy the delicious fruits of your labor with your group!

Adapting Recipes to Outdoor Cooking Conditions

Got a plan to head into the Great Outdoors? It's super important to adapt your recipes for cooking outside. You want meals that are tasty and fun to make. Let's dive into some tips on tailoring your recipes for an open fire:

Simplify Ingredients:

First off, go for recipes with fewer ingredients. Minimal prep is key! Stick to pantry staples like grains, canned goods, and dried spices. They're lightweight and pack easily.

Think about what you might find in the wild. Choose recipes using local or foraged items.

Embrace One-Pot Meals:

One-pot meals are fantastic for camping. They simplify cooking and cut down on cleanup.

Look for recipes where you can toss in veggies, carbs, and proteins all in one pot or skillet. Stir-fries, stews—these are excellent options.

Adjust Cooking Times:

Cooking outside is different from the kitchen. Factors like wind, temperature, and fire intensity affect cooking times.

Be ready to tweak timings. Watch your food carefully and rely on your senses rather than just a timer.

Experiment with Foraged Foods:

Nature offers lots of wild edibles and seasonal goodies. Using them makes your meals fresh and exciting.

Learn to spot edible plants, mushrooms, and berries around your campsite. Add them as main ingredients or fun complements.

Adapt Cooking Techniques:

Change up your techniques for the outdoors. If grilling meat over flames, maybe use indirect heat to avoid burning.

Try out food preservation methods like smoking or drying. This can prolong the life of perishables and add rich flavors.

CONSTRUCTING AND SUSTAINING OUTDOOR COOKING FIRES

Cooking without your usual kitchen gadgets is different from cooking indoors. Picking the right firewood, handling the heat—it all matters (a lot). It's a longer process too, but it's worth it! Let's dive into the must-know techniques and tips for building & keeping outdoor cooking fires.

Safety Tips for Choosing & Preparing Firewood

Safety First! When it comes to fire, being safe is super important. Picking the right firewood can make cooking easier and safer. Here's how to choose the best stuff:

- **Choosing Firewood:** Go for dry, seasoned hardwoods like oak, maple, or hickory. Avoid green or sappy woods—they make too much smoke and sticky sap. That messes with the taste of your food and doesn't keep a reliable fire.
- **Preparing Firewood:** Get different sizes of wood—from big logs to tiny kindling. Use a strong axe or saw to chop them up into easy-to-handle pieces. Make sure they're dry because wet wood doesn't burn well.

Techniques for Different Types of Outdoor Cooking Fires

Being good at making different kinds of fires can really boost your outdoor cooking game. Here's how:

- **Campfires:** Campfires are classic! They give you warmth, light, and a steady heat source for cooking. Clear out any flammable stuff around the area you want to use. Put large logs or rocks in a circle to keep the fire in place. Pile up smaller kindling and tinder in the middle and light it with matches or a fire starter. Slowly add bigger wood pieces as the fire grows. Remember Chapter 5—it's all about building a safe and long-lasting fire!
- **Rocket Stoves:** Rocket stoves are cool because they're efficient and portable! Stack bricks or rocks vertically, leaving a small gap for air to flow through. On top, put a grate or metal surface for your pots and pans. Feed small sticks or twigs into the stove's opening. The vertical stack makes a powerful draft that pumps out strong heat with very little smoke.

Strategies for Keeping & Controlling Fire Temperature

Getting just the right heat is key for great outdoor cooking. Here's what to do:

- **Fuel Management:** Change up the size and layout of your firewood to control how hot it gets. Smaller pieces mean bigger flames (and more heat). Covering parts of the fire reduces airflow and lowers the temperature.
- **Airflow Control:** Move your pots closer or farther from the flames to tweak the heat level. Using adjustable grates helps too! Putting rocks or logs around the fire can also guide airflow and trap heat.
- **Monitoring:** Keep an eye on that temperature! Use a grill thermometer or watch how colorful and intense the flames are. Adjust as needed to keep things cooking just right.

OVERVIEW OF SOLAR COOKING TECHNIQUES

Ever thought about using the sun to cook? It's a pretty cool and eco-friendly way to make your meals. Solar cooking taps into the sun's power, giving you a neat and sustainable solution for preparing food. By using the sun's energy, you can cook efficiently and in an environmentally friendly way. So, let's dig into what solar cooking is all about and explore its principles, benefits, and practical uses.

Explanation of Solar Cooking Principles & Benefits

Solar cooking works on simple yet effective principles: **solar radiation** and **thermal energy transfer**. When sunlight hits a solar cooker, it's absorbed and turned into heat. This heat then cooks your food. The beauty of it? No need for gas or charcoal—it's all about harnessing the sun.

Why go solar with your cooking? Here are some perks:

- **Sustainability:** You can whip up veggies, carbs, and proteins all in one pan or pot. Think stir-fries, stews, soups—those good one-pot wonders.
- **Cost-effectiveness:** After you build it, a solar cooker doesn't need more fuel. It's an economical choice for outdoor cooking fans.
- **Healthier Cooking:** Slow and steady wins the race! Solar cooking retains more nutrients in your food, resulting in healthier and tastier meals.
- **Versatility:** Wherever there's sunlight, you can use a solar cooker. From backyard BBQs to far-off camping trips—it's reliable in many settings.

Step-by-Step Instructions for Constructing Simple Solar Ovens

Making a solar oven can be fun and rewarding, using basic materials and tools. Here's how you do it:

Gather Materials:

- **Cardboard Box:** Find a sturdy box that's a good size for your oven.
- **Aluminum Foil:** Line the inside of the box with foil (shiny side facing inward) to reflect sunlight.
- **Transparent Lid:** Use glass or clear plastic as a lid to let sunlight in but keep heat trapped.
- **Insulating Materials:** Grab things like crumpled newspaper or foam board to insulate between the box's walls.
- **Dark-Colored Pot or Pan:** A dark pot with a lid helps absorb heat better.
- **Additional Tools:** Scissors, adhesive tape/glue—whatever helps secure everything in place.

Prepare the Box:

1. Open up the box flat and remove any flaps for an even surface.
2. Line it with foil (shiny side inside) and secure it with tape or glue.

Create Insulation:

- Pack insulating materials like newspaper tightly around the inside walls.
- This layer helps keep heat inside for more effective cooking.

Construct the Lid:

- Cut your glass/plastic sheet to fit snugly on top as the lid.
- Secure it with tape or glue so it stays in place during use.

Position the Cooking Vessel:

- Place your dark-colored pot or pan in the center of your oven.
- Make sure it's stable and level to avoid spills.

Maximize Sun Exposure:

- Put your oven where it'll catch plenty of sun—on the patio, in the backyard, or at a campsite.
- Adjust its angle to follow the sun's path during the day. Avoid shaded or obstructed spots.

Tips for Maximizing Solar Cooking Efficiency

Want to get the best out of your solar cooker? Try these tips:

- **Optimize Placement:** Always put it where sunlight is unobstructed.
- **Adjust Cooking Times:** Pay attention to weather changes; sunny days cook faster than cloudy ones.
- **Use Reflective Surfaces:** Add mirrors or more foil around your cooker to boost sunlight direction into it.
- **Practice Patience:** Solar cooking is slow but worth it! Give enough time for thorough cooking, especially at lower temperatures.

CREATIVE OUTDOOR COOKING: NUTRITIOUS MEAL IDEAS FOR THE WILDERNESS

When you're in a survival situation it's crucial to have meals that are easy to make, require minimal ingredients, and provide the necessary nutrition to keep you going. Here are some straightforward recipes that fit the bill.

Breakfast

1. **Quick Egg Scramble**
    - **Ingredients:** Powdered eggs, dried spinach, salt, pepper.
    - **Instructions:**

1. Rehydrate powdered eggs and spinach with water.
2. Cook in a pan over a portable stove or campfire until eggs are set.
3. Season with salt and pepper and serve.

2. **Instant Oatmeal**
   - **Ingredients:** Instant oats, dried fruit, nuts, honey.
   - **Instructions:**
     1. Add hot water to instant oats in a bowl.
     2. Stir in dried fruit and nuts.
     3. Drizzle with honey and eat warm.

3. **Protein Pancakes**
   - **Ingredients:** Pancake mix, water, powdered milk, protein powder.
   - **Instructions:**
     1. Mix pancake mix, water, powdered milk, and protein powder.
     2. Cook on a hot skillet or pan until golden brown on both sides.
     3. Serve plain or with any available syrup.

4. **Breakfast Wrap**
   - **Ingredients:** Tortillas, canned beans, cheese, hot sauce.
   - **Instructions:**
     1. Warm tortillas on a skillet or directly over a fire.
     2. Fill with canned beans and cheese.
     3. Add hot sauce if available, then roll up and eat.

5. **Survival Porridge**
   - **Ingredients:** Instant oats, powdered milk, sugar, water.
   - **Instructions:**
     1. Combine oats, powdered milk, and sugar in a bowl.
     2. Add hot water, stir until thickened, and eat immediately.

Lunch

1. **Canned Tuna Wrap**
   - **Ingredients:** Tortillas, canned tuna, mayo or oil, salt, pepper.
   - **Instructions:**
     1. Mix canned tuna with a bit of mayo or oil.
     2. Season with salt and pepper.
     3. Spread onto a tortilla, roll up, and eat.

2. **Rice and Beans**
   - **Ingredients:** Instant rice, canned beans, seasoning packet.
   - **Instructions:**
     1. Cook instant rice according to package instructions.
     2. Heat canned beans and mix with rice.
     3. Season with a seasoning packet and serve.

3. **Emergency Salad**
   - **Ingredients:** Canned vegetables, canned meat, oil, vinegar, seasoning.

- **Instructions:**
    1. Drain canned vegetables and meat.
    2. Toss together with oil, vinegar, and seasoning.
    3. Serve cold.
4. **Simple Pasta**
    - **Ingredients:** Instant pasta, canned tomatoes, canned meat (optional).
    - **Instructions:**
        1. Cook instant pasta according to instructions.
        2. Stir in canned tomatoes and meat if available.
        3. Heat through and serve.
5. **Peanut Butter Sandwich**
    - **Ingredients:** Bread or crackers, peanut butter, honey.
    - **Instructions:**
        1. Spread peanut butter onto bread or crackers.
        2. Drizzle with honey if available.
        3. Close sandwich or stack crackers and eat.

Dinner

1. **One-Pot Stew**
    - **Ingredients:** Canned meat, canned vegetables, instant rice or potatoes, seasoning.
    - **Instructions:**
        1. Combine canned meat and vegetables in a pot.
        2. Add instant rice or potatoes and enough water to cover.
        3. Simmer until heated through and serve.
2. **Foil Packet Meal**
    - **Ingredients:** Canned meat, canned potatoes, seasoning, aluminum foil.
    - **Instructions:**
        1. Place canned meat and potatoes on a sheet of aluminum foil.
        2. Season as desired, then wrap tightly in the foil.
        3. Cook over a fire or portable stove for 15-20 minutes and serve.
3. **Simple Chili**
    - **Ingredients:** Canned beans, canned tomatoes, chili seasoning, canned meat (optional).
    - **Instructions:**
        1. Combine all ingredients in a pot.
        2. Heat over a fire or portable stove until bubbling.
        3. Serve hot.
4. **Survival Pasta**
    - **Ingredients:** Instant pasta, canned sauce, canned vegetables.
    - **Instructions:**
        1. Cook pasta according to package instructions.
        2. Stir in canned sauce and vegetables.
        3. Heat until warm and serve.

5. **Grilled Cheese**
    - **Ingredients:** Bread, cheese slices, butter or oil.
    - **Instructions:**
        1. Butter one side of each bread slice.
        2. Place cheese between two slices, buttered side out.
        3. Cook in a pan until golden brown on both sides and cheese is melted.

Snacks

1. **Trail Mix**
    - **Ingredients:** Nuts, dried fruit, chocolate chips.
    - **Instructions:**
        1. Mix all ingredients together in a resealable bag.
        2. Store for a quick energy boost.
2. **Peanut Butter Crackers**
    - **Ingredients:** Crackers, peanut butter.
    - **Instructions:**
        1. Spread peanut butter on crackers.
        2. Top with another cracker and eat.
3. **Energy Bars**
    - **Ingredients:** Rolled oats, peanut butter, honey, dried fruit.
    - **Instructions:**
        1. Mix all ingredients together.
        2. Press into a pan or mold and let set.
        3. Cut into bars and store for quick snacks.
4. **Campfire Popcorn**
    - **Ingredients:** Popcorn kernels, oil, salt.
    - **Instructions:**
        1. Heat oil in a pot over the fire.
        2. Add popcorn kernels and cover.
        3. Shake occasionally until popping slows, then remove from heat and salt.
5. **Canned Fruit and Nuts**
    - **Ingredients:** Canned fruit, nuts.
    - **Instructions:**
        1. Open canned fruit and drain excess liquid.
        2. Mix with nuts and eat as a refreshing snack.

CRUCIAL GUIDELINES FOR SAFE AND HYGIENIC OUTDOOR COOKING

When you're out in the wild, keeping up with food safety and hygiene is just as important as when you're indoors. It's key to avoid foodborne illnesses and ensure your efforts pay off. Otherwise, things could take a turn for the worse. So, let's dive into what you need to do to stay safe with food in the wilderness.

Importance of Practicing Proper Food Safety & Hygiene

- **Purpose:** Sticking to proper food safety & hygiene practices is essential to prevent foodborne illnesses. It keeps you and your fellow adventurers well and happy.
- **Enthusiastic Explanation:** When you're in the wild, proper food safety and hygiene is your first defense against getting sick from food. This is super crucial because medical help might not be nearby. By following these tips, you can totally enjoy your outdoor cooking with peace of mind.

10. Solar Power

GETTING STARTED WITH SOLAR POWER DIY PROJECTS

Solar power offers a dependable energy source separate from the grid, making it perfect for those wanting to be self-reliant preppers. This chapter dives into the basics of DIY solar power systems, especially focusing on customized off-grid setups.

Why Solar Power Matters for Preppers

Having access to a powerful and renewable energy source like solar power is truly a boon for preppers. Whether prepping for natural disasters, economic woes, or societal turmoil, solar power ensures constant electricity when usual sources falter. It's indispensable, and here's why:

- **Self-Sufficiency:** Solar power empowers preppers to break free from reliance on centralized utilities. It fosters self-sufficiency and resilience in tough times.
- **Renewable and Sustainable:** Unlike fossil fuels, solar energy is abundant, clean, and renewable. It minimizes environmental impact while ensuring long-term sustainability.
- **Cost-Effective:** DIY solar power systems can lead to significant savings by slashing or eliminating monthly electricity bills.
- **Versatility:** These systems can be tailored to different needs—from powering essential appliances in off-grid cabins to offering backup power for urban homes during outages.

Understanding Off-Grid Solar Systems

Off-grid solar systems operate without utility connections, storing extra energy in batteries for cloudy days. Here's a quick rundown of their components and functionality:

- **Solar Panels:** The backbone of any system, these panels convert sunlight into electricity through photovoltaic cells. Choose high-efficiency models that maximize energy even in low light.
- **Charge Controllers:** These regulate electricity flow from panels to batteries, preventing overcharging and extending battery lifespan. Opt for Maximum Power Point Tracking controllers—they optimize energy conversion for better efficiency.
- **Batteries:** These store excess energy generated by panels for use at night or during low sunlight. Lithium iron phosphate (LiFePO4) batteries, like Battleborn Batteries, are ideal due to their high density, long life, and maintenance-free operation.
- **Inverters:** They convert direct current (DC) stored in batteries into alternating current (AC) used by household appliances. Victron Quattro inverters offer reliable performance and seamless integration with solar systems.
- **Monitoring and Maintenance:** Longevity and optimal operation depend on regular maintenance and monitoring. Use devices like the Victron Lynx Distributor and Lynx Shunt to track energy production and battery health. Perform checks on connections, clean panels periodically, and ensure ventilation to avoid overheating.

SELECTING THE OPTIMAL ELEMENTS

Embracing the adventure of setting up your own DIY solar power system is truly exciting! Selecting the right components is key to ensuring it's efficient, reliable, and long-lasting. Let's dive into the essential parts of a solar power system and provide you with some practical tips on picking the best ones for your needs.

Solar Panels

Solar panels are the heart and soul of any solar power setup; they transform sunlight into electricity. When picking out solar panels, keep these factors in mind:

- **Efficiency:** If you want to get the most electricity from a small area, look for panels that have high efficiency ratings.
- **Durability:** Choose weather-resistant materials and sturdy construction, especially if you expect rough weather.
- **Type:** Depending on your space and budget, pick between monocrystalline, polycrystalline, or thin-film solar panels.
- **Warranty:** It's wise to select panels that come with a warranty (ideally covering at least 25 years) to protect your investment.

To figure out how many panels you need, calculate the total wattage needed for your appliances and devices—consider both current and future energy usage. Use online calculators or ask a professional for help in deciding the optimal number of panels.

Batteries

Batteries store extra solar energy made during the day for those cloudy periods or nighttime use. Here's what to consider when picking batteries:

- **Battery Type:** Choose deep-cycle batteries like lithium-ion or lead-acid, specially designed for renewable energy systems.
- **Capacity:** Determine the needed battery capacity (measured in ampere-hours, Ah) based on your energy use patterns and the level of independence you want.
- **Cycle Life:** Go for batteries with a high cycle life—these can handle lots of charge-discharge cycles without wearing out.
- **Maintenance:** Consider maintenance needs. Maintenance-free batteries can simplify life a lot!

Ensure your battery bank's capacity meets or—better yet—exceeds your household's total energy consumption so you don't face shortages during times of low solar input.

Charge Controllers

Charge controllers are essential gadgets that maximize battery life by controlling the electricity flow from solar panels to batteries and preventing overcharging. Consider these factors when choosing charge controllers:

- **Voltage Compatibility:** Select one that matches both your battery bank's and panel array's voltage.
- **Type:** Based on your system's voltage and efficiency needs, choose between Maximum Power Point Tracking (MPPT) and Pulse Width Modulation (PWM) controllers.
- **Maximum Current:** Ensure its maximum current rating is higher than what your solar panels produce.

Right-sized charge controllers ensure energy conversion is efficient and keep your batteries safe from overcharging or heavy discharge.

Inverters

Inverters convert DC electricity stored in batteries into AC power usable by household gadgets. Look at these factors when selecting inverters:

- **Power Rating:** Pick inverters with enough power output to cover the peak demands of your appliances and devices.
- **Waveform:** Decide between pure sine wave and modified sine wave inverters based on how sensitive your electronics are.
- **Efficiency:** Seek high-efficiency inverters to reduce energy losses during conversion.
- **Safety Features:** Go for inverters equipped with safeguards against overvoltage, overload, and short circuits.

Quality inverters ensure steady and reliable power delivery to keep your electronic devices running smoothly.

SETTING UP THE SYSTEM

Installation Process

Let's get started on your journey to energy independence! Here's a handy, step-by-step guide to properly set up your solar panels—from mounting them to wiring for top performance.

Mounting Solar Panels

1. **Find the Perfect Spot:**
   Seek out a spot with the most sunlight throughout the day. A south-facing location is usually best, catching all that sweet sunlight.

2. **Prep the Surface:**
   Make certain your mounting surface is stable, clean, and level. Clear away any clutter or anything that might block the sun's rays.

3. **Install Mounting Brackets:**
   Attach the brackets to the surface with screws or bolts. Ensure they're firmly in place.

4. **Place Your Panels:**
   Gently lift and place each panel onto the brackets. Align them well to soak up sunlight fully. Use a level to ensure they are perfectly horizontal.

5. **Secure Those Panels:**
   Once in position, lock them down with nuts or clamps. Make sure they're tightly secured so they can brave any weather conditions.

6. **Check Alignment Again:**
   Double-check your panels' direction and angle to ensure they're perfectly set up for grabbing sunshine.

Wiring the Solar Panels

1. **Gather Your Tools & Materials:**
   Before you begin wiring, collect everything you'll need—solar cables, MC4 connectors, wire strippers, and a crimping tool.

2. **Connect Panels in Series or Parallel:**
   Decide whether you'll connect in series (for more voltage) or parallel (to boost current capacity).

   - **Series Connection Setup:**
     For higher voltage, connect the positive terminal of one panel to the negative terminal of another, repeating until done. Secure connections with MC4 connectors.

   - **Parallel Connection Setup:**
     For maintaining voltage but increasing current capacity, just connect all positive terminals together and then do the same for negatives.

3. **Terminate Wiring:**
   After connecting, attach MC4 connectors to your cable ends for a secure and weatherproof finish.

4. **Route Cables Carefully:**
   Route solar cables from panels to the charge controller thoughtfully—protect them from damage and secure them well.

5. **Hook Up to Charge Controller:**
   Follow the manufacturer's instructions closely when attaching wires to the charge controller terminals. Make sure connections are tight and polarity is correct to avoid issues.

6. **Test Everything:**
   Before you wrap up, thoroughly test your setup. Ensure everything is wired properly and working as expected.

WIRING BATTERIES, CHARGE REGULATORS, AND POWER INVERTERS

Battery Connections:

Got all your lithium iron phosphate batteries ready? They're light, durable, and need hardly any maintenance compared to those old lead-acid ones. A perfect pick if you want something that'll last and be trusty.

Step 1: Check Insulation & Mounting

First thing's first—check the insulation and mounting. You gotta make sure the batteries are insulated properly and fixed tight in your system. If they aren't, you could face some serious safety issues, and your solar setup might not work as well as it could.

Step 2: Link the Batteries

Now, let's link those batteries! Use top-notch battery cables to hook them up according to what your system needs voltage-wise. Whether it's a 24V or 48V setup, you'll wire them differently.

- **Series Connection:** Going for 48 volts? Connect the positive side of one battery to the negative of the next one. Keep doing this until all batteries are hooked up. This doubles the voltage but keeps the capacity the same.

- **Parallel Connection:** If you want 24 volts, connect all positive terminals together and all negative terminals together. This way, you keep the same voltage but boost the capacity—great if you need more energy storage.

Step 3: Secure Connections

Use cable lugs and crimping tools for secure connections. Cable lugs make a strong link between battery terminals and cables, cutting down resistance and boosting power transfer efficiency. Proper crimping locks everything in place tightly, reducing electrical faults and keeping things running smoothly.

Step 4: Double-Check All Connections

Inspect every link carefully to make sure everything's tight and consistent with polarity. Loose or wrong connections can cause drops in voltage, overheating, and maybe even system failures. Take your time here—it's worth it!

Charge Controller Installation:

Step 1: Mount the Charge Controller

Pick a spot that balances good ventilation with easy access. Plus, it should be close enough to both the battery bank and solar panels. Make sure whatever you're mounting on is sturdy and can hold the weight.

Step 2: Connect Solar Panel Array

Get wires of the right size and connect your solar panel array's positive and negative ends to matching slots on the charge controller. Follow what the manufacturer says about wire gauge, voltage, and current ratings so you don't overload things.

Step 3: Grounding

Grounding's important too! It helps manage risks with electrical faults and keeps operations safe by giving extra current a place to go.

Step 4: Neaten Up the Wiring

Neaten up your wiring with zip ties or cable clips to minimize clutter. This prevents damage or interference and also makes troubleshooting easier later on—not to mention it looks way better!

Inverter Integration:

Step 1: Position the Inverter

Find a clean, dry spot that has good airflow for cooling things down during operation. Proper ventilation stops overheating, which can mess with performance or shorten inverter life.

Step 2: Hook Up to Battery Bank

Use heavy-duty cables for this part. Connect the battery bank's positive and negative ends to the matching slots on the inverter securely so they handle the load well without big voltage drops.

Step 3: Configure Inverter Settings

Set input/output voltages according to the manufacturer's recommendations for compatibility and best performance.

Step 4: Test Everything

Finally—test everything out! Turn on the inverters after hooking them up and check their performance closely. Are they syncing properly? Are they delivering steady AC power to home loads? Watch voltage levels, waveform quality, and system stability until you're satisfied things are running just as they should!

ONGOING SYSTEM CHECKS & UPKEEP

Importance of Keeping an Eye on System Performance

Ensuring that your DIY solar power system operates efficiently is crucial. Monitoring its performance allows you to catch problems early and fix them before they escalate. Installing a solar monitoring system can be incredibly beneficial. It provides real-time updates on key metrics such as solar panel output, battery levels, and energy use. With this information, you can quickly identify any anomalies and take the necessary actions.

Regular checks enable you to fine-tune your energy production. You can spot issues like shading from trees or dirt on the panels that may be reducing efficiency. Addressing these issues can enhance your system's energy output.

Basic Maintenance Tasks for Top Performance

While DIY solar power systems generally require minimal upkeep, some tasks are essential to keep things running smoothly.

- **Inspect Solar Panels:** Regularly examine your panels for damage or dirt buildup. Clean them with a soft brush or cloth to maintain efficiency.
- **Check Battery Levels:** Monitor your battery charge levels to ensure they are functioning properly. Significant drops in voltage could indicate a problem.
- **Inspect Wiring & Connections:** Check all wires and connections for signs of wear or corrosion. Repair any damaged wires and tighten loose connections to prevent electrical issues.
- **Test Inverter Operation:** Regularly test your inverters to ensure they are efficiently converting DC power into usable AC power.
- **Run System Diagnostics:** Utilize your solar monitoring system to perform routine diagnostics on your entire setup to identify issues before they lead to bigger problems.

Safety First!

Safety is paramount, no matter what part of the system you're working on:

- **Wear the Right Safety Gear:** Always use insulated gloves, safety goggles, and non-conductive shoes before starting any work.
- **Disconnect Power Sources:** Always disconnect power sources to prevent accidents. Turn off the main breaker and disconnect the batteries first.
- **Test for Live Wires:** Use a non-contact voltage tester before touching any components.
- **Work in Dry Conditions:** Avoid working in wet conditions; ensure all tools and gear are dry as well.
- **Ground Components Properly:** Ensure that all components—solar panels, batteries, inverters—are properly grounded.
- **Protect Against Overloading:** Install circuit breakers or fuses on all electrical circuits to prevent overloading.

Troubleshooting Common Issues

Even with proper care, issues can arise. Here are some troubleshooting tips:

- **Low Power Output:** If energy output is low, first check the panels for damage or shading. Clean any debris and inspect the wiring connections.
- **Battery Issues:** If batteries aren't holding a charge or die quickly, check for tightness and corrosion at the battery connections. Ensure they are correctly sized and charged properly by the charge controller.
- **Inverter Problems:** If there are inverter issues, check for loose or damaged input/output wires. Ensure it's receiving power from the batteries and is set up correctly for your system voltage. Consult the manual or tech support if needed.
- **Monitoring System Performance:** Continuously monitor your system using the tools available or through manual checks.
- **Seek Professional Help If Needed:** Don't hesitate to call a professional if you encounter persistent problems that you can't resolve on your own. They'll ensure your system functions safely and efficiently.

You've got this! Keep monitoring regularly and addressing small issues promptly to ensure your DIY solar power system operates optimally for years to come.

PROJECT 12: SET UP YOUR OWN SOLAR PANEL SYSTEM

You don't have to spend a small fortune to have a small solar panel system. With the right tools, materials, and attitude, you can create your own system so that you have power, even during a disaster.

Materials and Preparation

Before you start, you will need the following materials:

- Solar panel starter system
- 100ah 12v SLA Battery
- Power Inverter (500w)
- Two inverter cables (make sure that the inverter cables you get will work with your inverter)
- Adjustable wrench/Pliers
- Plywood
- Flathead screwdriver

You will want to make sure you have enough room for your work. The larger the panels, the more space you will need.

Steps

1. Connect the charge controller to the battery.
    a. Mount the charge controller to plywood, which will be near the battery.
    b. Remove the insulation on the wire.
    c. Look for the Battery Icon on the controller with a negative (-) sign, then insert the wire into the space.
    d. Use the flathead screwdriver to tighten it.
    e. Gently tug on the wire to make sure it is secure (won't fall out).
    f. Run the wire to the negative (-) terminal on the battery. The negative terminal will either have a negative (-) symbol or it will be black.
    g. Repeat steps 1b to 1f for the positive (+) wire, routing it from the positive symbol on the controller to the battery's positive (+) terminal. The terminal will either have a positive (+) symbol or it will be red.
    h. Check that the green light showed on the charger. This means that the wires are properly attached.
2. Setup the first solar panel and connect it to the charge controller.
    a. Plug the negative (-) cable to the solar panel. The negative cable is all black.
    b. Plug the other end of the negative (-) cable into the charge controller slot for the negative (-) cable.
        i. Unscrew the slot.
        ii. Insert the cable.
        iii. Screw the slot back in, making sure it is secure.

    c. Repeat steps 3a and 3b with the positive (+) wire. For the solar panel connect, the positive (+) is red.
3. Take the panel outside and test it to make sure it charge. You need enough sun to charge.
    a. Two of the lights should be a solid green.
    b. One light (PV indicator) should be flashing green t show it is charging.

Once the system is charged, you'll want to be able to use the energy stored in the battery.

4. Remove one of the solar power cables from the controller, then the same wire from the battery.
5. Connect the positive (+, red) cable and screw it down on the inverter, then the negative (-).
ALWAYS connect both cables to the inverter first.
6. Connect the wires to the corresponding terminals on the battery.

You can have several panels, each with their own battery if you need more power.

PROJECT 13: BUILD A SOLAR FOOD DEHYDRATOR IN YOUR BACKYARD

This project is far simpler than it sounds (and certainly easier than a solar panel system), but it will take time to finish as you will be making it from scratch. When you finish, you will have a safe, reliable way to dehydrator your food for safe storage for later. Best of all, it won't cost nearly as much as buying a system.

Materials and Preparation

Before you start, you will need the following materials:

- 4 Plywood that is at least 2.5 long and 4 wide
- 2x2 wood that is 10 ft
- Window that fits into the door
- Heat absorbent shelf
- Screen
- Cloth/stretchable material
- Door fastener (could be as simple as a hook and string)
- 2 hinges
- Pack of screws
- Staples
- Thermometer
- Caulk
- A saw to cut the plywood

Use the following image to help you purchase the right wood for your dehydrator.

Dehydration Unit Sketch

Steps

Use the following steps to make your own dehydration unit.

1. Cut out the following using the dimensions provided under the materials.
    a. The top
    b. The two sides
    c. The bottom
    d. The back
    e. The legs and braces
2. Create the base.
    a. Cut notches in the side of the bottom to fit the four legs.
    b. Attach the legs to the bottom of the unit.
    c. Add the bracers to the legs, about halfway up from the ground.
    d. Attach side pieces, anchoring the pieces by screwing them into the bottom.
3. Size and install the heat absorbent shelf on the bottom.
4. Stretch the cloth or stretchy material over the frame.
5. Attach the screen.
6. Install the window to the front panel., using caulk to fill any gaps.
7. Install the thermometer inside near the drying screen rack.

Finished Dehydration Unit

You will need to store the unit somewhere that is sunny. Dehydration requires the food to be exposed to between 100 and 140 degrees F, depending on what the food is. That means you will need it to get enough heat to dry the food.

PROJECT 14: CREATE A SOLAR OVEN FOR EASY OUTDOOR COOKING

During a crisis, you'll want to be able to cook, and you may not have power to do that. You will probably have a lot to do, so having an oven, instead of just an open fire, can give you a level of freedom to get stuff done while having food being prepared. What's really great about this one is that is a science project for kids, so you know that it is simpler than many of the other projects.

Materials and Preparation

Before you start, you will need the following materials:

- A cardboard box (this could even be a pizza box)
- Black construction paper
- Newspapers
- Aluminum foil
- Plastic wrap
- Scissors
- Clear tape
- Ruler
- Thermometer

Steps

1. Cut a square flap in the box lid. This means cutting three sides, then leaving the fourth as a kind of hinge for the flap.
2. Add aluminum foil to the inside part of the flap.

The Flap with Aluminum Foil

3. Add plastic wrap over the box with the flap up. This means that the flap will up, not closed. Tape the plastic wrap down on the sides of the box so that it stays in place.
4. Add the construction paper to the inside bottom of the box.
5. Roll up several pieces of newspaper and line all four walls inside the box. Tape them in place so they don't move.

Now you have a fire free oven that you can use to cook foods. You will want to keep it in direct sunlight to cook. Something like a small pan, such as a pie pan, will ensure you don't get the bottom dirty.

Note: Put the thermometer in the box before you close the lid to cook food. This will let you know that it got hot enough to cook the food.

11. Self-Defense

The heart of self-defense for preppers revolves around being physically ready & having a strong mental framework. This helps someone anticipate, evaluate, and handle potential threats well. Let's dive into the key details of self-defense. We'll highlight the prepper's mindset, basic self-defense principles, mental & physical readiness, situational awareness, and building a strong mental attitude for self-defense situations.

Self-defense starts with physical preparation. It's about strength and agility. Equally important is forming a mental framework to foresee issues before they arise.

Why do you need this? Understanding these nuances can make all the difference when faced with danger. The prepper's mindset includes staying calm & collected in stressful situations. It also means using foundational self-defense principles effectively.

GRASPING THE SURVIVALIST MENTALITY FOR PERSONAL PROTECTION

Self-defense is more than just learning how to protect yourself—it's about living a lifestyle focused on being prepared. Skills are just as important as having food and water stored. This mindset involves recognizing that tough times might come when society doesn't function as it should, and knowing that you'll need to rely on yourself for safety.

Overview of Self-Defense Principles and Ethics

Self-defense means you have the right to keep yourself safe from harm using reasonable force. Every person has the natural right to be safe. The rules of self-defense are based on principles of necessity, using no more force than required, and trying to avoid conflicts whenever possible. You must use force wisely, applying only what's needed to stop a threat.

Mental and Physical Preparation

Preparation requires both a strong body and a strong mind. Physically, it involves practicing self-defense techniques, understanding how the body works, and staying fit for quick responses. Mentally, it's about always being aware, staying calm under pressure, and having the will to act when danger approaches.

Situational Awareness and Threat Assessment

A key aspect of self-defense is being aware of your surroundings—spotting trouble before it starts. This skill involves constantly observing, reading body language that might indicate anger, and knowing your environment. By identifying threats early, you can avoid dangerous situations and prepare to respond appropriately.

Building a Strong Mental Attitude for Self-Defense

A strong mental attitude is essential. This involves developing confidence in your abilities, remaining calm when danger arises, and thinking through different scenarios in advance. Part of this mindset is understanding that self-defense situations can be emotionally challenging and preparing for how to handle your feelings after a tough encounter.

In self-defense, the focus is on preventing danger before it happens and de-escalating situations if they do arise. However, if you must defend yourself, it's crucial to understand the legal and moral implications of using force. Training should emphasize real-world techniques that are effective in various situations, with a strong emphasis on avoiding and escaping danger. Flexibility, quick thinking, and having a comprehensive plan for staying safe are central to prepping for self-defense.

GETTING STARTED WITH PERSONAL DEFENSE

Importance of Self-Defense Skills

Being able to defend yourself is really important. Imagine walking down a dark street or dealing with a scary situation at home. Knowing self-defense can make a huge difference.

- **Personal Safety:** The main goal of self-defense is to keep you and your loved ones safe. By learning how to defend yourself, you can handle tough situations better and reduce the risk of getting hurt.
- **Empowerment:** Learning self-defense makes you feel strong and confident. When you know you can protect yourself, you feel more secure and less scared in difficult times.
- **Preparedness:** Knowing self-defense means you can quickly react when in danger. It's about being ready before something bad happens. This way, you're taking control and keeping everyone safe.
- **Physical and Mental Health:** Practicing self-defense is good for both your body and mind. It builds strength, flexibility, and quickness. Plus, it helps you stay focused and disciplined. Overall, it makes you healthier and happier.

Overview of Self-Defense Martial Arts

There are many types of martial arts that help with self-defense. Each one has its own methods and principles. Here are some popular ones:

- **Krav Maga:** Originating from Israel, Krav Maga is very practical and used by military and police forces worldwide. It focuses on quick, decisive moves. You learn punches, kicks, elbows, and knee strikes—perfect for real-life situations.
- **Brazilian Jiu-Jitsu (BJJ):** BJJ is all about grappling and using technique instead of strength. It teaches you how to control someone on the ground using joint locks and chokeholds. It's great for close-up self-defense.
- **Muay Thai:** From Thailand, Muay Thai uses the fists, elbows, knees, and shins in fighting techniques. Known as the "Art of Eight Limbs," it's fantastic for striking powerfully and accurately. Plus, it boosts fitness and endurance.
- **Taekwondo:** A Korean martial art known for its fast movements and powerful kicks. It helps with agility and precision in attacks like kicks, punches, and blocks—useful against multiple attackers.

- **Boxing:** This old combat sport focuses on punching mechanics, footwork, and head movement. It's easy to learn and efficient for blocking hits while landing strong punches. Boxing improves hand-eye coordination and timing—great skills for defense.

Each martial art offers something special and can be tailored to what suits you best. Whether you're into striking or grappling—or maybe both—trying out different martial arts helps build a solid set of self-defense skills.

ESSENTIAL SURVIVAL METHODS

Getting good at core techniques is like making a blade super sharp—it's where all other skills come from. Krav Maga, BJJ, Muay Thai, Taekwondo, and Boxing each have their own style, offering lots of tools to be effective. Let's take a look at these disciplines to see their power and skills.

Krav Maga: Punches, Kicks, & Defensive Moves

Purpose: Krav Maga was made for real-world self-defense situations. It's all about quick and efficient actions to stop threats fast.

Punches & Kicks: Krav Maga teaches a variety of punches and kicks, focusing on powerful strikes to targets like the groin, throat, and eyes. You will practice palm strikes, elbow strikes, and knee strikes with great precision.

Defensive Moves: A big part of Krav Maga is simultaneously defending and attacking. Practice blocking and countering smoothly, using natural movements to dodge hits while striking back hard.

Brazilian Jiu-Jitsu: Grappling and Ground Fighting

Purpose: BJJ helps you defeat bigger opponents using leverage and technique. Ground fighting is key, so even when you're on the ground, you can still defend yourself effectively.

Grappling Techniques: BJJ focuses on control and submission holds, using joint locks and chokeholds to subdue opponents. Learn basic takedowns and maintain control to manage the fight.

Ground Fighting Skills: Use the ground to your advantage with BJJ techniques to neutralize threats and secure dominant positions. Work on transitioning between positions and applying submission holds to be effective from any angle.

Muay Thai: The Art of Eight Limbs

Purpose: Muay Thai is known as the "Art of Eight Limbs" because it uses fists, elbows, knees, and shins as weapons. It emphasizes powerful strikes and constant pressure.

Striking Techniques: Develop perfect striking skills by practicing punches, elbows, kicks, and knees consistently. Focus on generating power from your hips and delivering hard-hitting strikes.

Clinch Work: Muay Thai excels in the clinch, allowing for close-range knee strikes and elbows. Master the clinch position to control opponents while inflicting damage with knees and elbows.

Taekwondo: Kicks and Strikes for Self-Defense

Purpose: Taekwondo is famous for its incredible kicks, offering a wide range of striking options for self-defense. It emphasizes speed, flexibility, and precise execution.

Kicking Skills: Taekwondo fighters have lightning-fast kicks that can keep opponents at bay. Train diligently to perfect kicks like the front kick, roundhouse kick, and sidekick for speed and accuracy.

Striking Techniques: In addition to kicks, Taekwondo also incorporates hand strikes and blocks. Practice punches and palm strikes, combining them with kicks for optimal effectiveness.

Boxing: Basic Punches & Footwork

Purpose: Boxing, also known as the "sweet science," focuses on precise punches and quick footwork to outmaneuver opponents. It builds discipline, fitness, and mental toughness.

Punching Fundamentals: Boxing has four main punches—jab, cross, hook, and uppercut. Drill these relentlessly, focusing on form, timing, and power transfer for maximum punch impact.

Footwork Skills: Excellent footwork is crucial for boxers, allowing them to dodge hits while creating angles to strike. Practice footwork drills to develop agility, balance, and control of distance in fights.

TACTICAL VIGILANCE

This section focuses on two crucial aspects of strategic awareness: **Situational Awareness** and **Escape and Evasion Techniques**.

Situational Awareness and Threat Assessment

Mastering situational awareness involves sharpening your senses to detect subtle signs of danger and opportunity around you. Here's how you can develop this vital skill:

- **Stay Alert:** Train yourself to always be vigilant, constantly scanning your surroundings for any potential threats or hazards.
- **Observe Cues:** Pay attention to anything unusual in your environment, such as strange behavior or unexpected changes.
- **Analyze Patterns:** Look for recurring behaviors or events. These could be indicators of a potential threat.
- **Trust Your Instincts:** Your gut feeling is often an excellent early warning system. If something feels off, trust it and take necessary precautions.
- **Practice Mindfulness:** Make mindfulness a habit. Stay present and avoid distractions that could compromise your awareness.

By incorporating these practices into your daily routine, you'll develop a strong sense of situational awareness that can be beneficial in everyday situations, not just during emergencies.

Escape and Evasion Techniques

Escape and evasion skills are essential components of self-defense, often prioritized over direct confrontation. Here are some key techniques to master:

- **Route Planning:** Always have multiple escape routes planned in advance, both for your current location and places you frequently visit.
- **Stealth Movement:** Learn to move quietly and inconspicuously, reducing your visibility and avoiding detection by potential threats.
- **Camouflage:** Use natural surroundings to blend in with your environment, making it difficult for adversaries to spot you.
- **Evasion Tactics:** If confronted by a threat, use evasion tactics such as zigzagging or seeking cover to disrupt your pursuer's line of sight.
- **Countertracking:** Be mindful of leaving tracks or traces that could reveal your movements. Practice countertracking techniques to minimize your footprint.
- **Urban Evasion:** In urban environments, utilize alleys, rooftops, and other hidden paths to navigate without attracting attention.

By adhering to these strategies, you'll be prepared, alert, and ready for whatever comes your way.

UNCONVENTIONAL SELF-DEFENSE TOOLS

Sometimes, in a pinch, you gotta think outside the box. Firearms and traditional weapons are great, but what if you need something more subtle or creative? That's where improvised weapons come in handy. Using everyday objects for defense is all about understanding impact and edged weapons. Let's explore how to turn stuff around you into makeshift weapons.

Everyday Objects as Defensive Tools

Imagine you're in a survival situation. Quick thinking and using what's at hand can make all the difference. Ordinary things can become powerful defensive tools with a bit of cleverness. Here's how you can turn common items into protective gear:

- **Flashlights:** Got a sturdy flashlight? Great! Its solid build and heavy end make it perfect for self-defense. Swing it to fend off anyone causing trouble.
- **Keys:** Your keychain isn't just for unlocking doors. Hold a key between your fingers, and you've got a handy tool to make your punches a lot stronger.
- **Belts:** A strong belt does more than hold up your pants. It can whip or even restrain someone, giving you time to get away or fight back.

- **Pens and Pencils:** Simple but effective. Aim for soft spots like the eyes or throat to incapacitate an assailant with these everyday writing tools.
- **Umbrellas:** Umbrellas aren't just for rain! Use one to strike, block, or keep an attacker at bay. The handle gives you extra reach, while the canopy confuses or catches them off guard.

Impact and Edged Weapons

Knowing how to use impact and edged weapons makes a huge difference in self-defense. Here's a look at each type and how they work:

- **Impact Weapons:** These rely on blunt force to stop or deter someone. Some common choices are:
    - **Baseball Bats:** A go-to for home defense, baseball bats are easy to use and hit hard. Aim for big muscles or the head for best results.
    - **Crowbars:** Heavy and made of steel, crowbars deliver serious blows. They can also be used for other tasks like prying open doors in an emergency.
    - **Brass Knuckles:** Though controversial, brass knuckles pack a punch (literally). They make your fists much more potent.
- **Edged Weapons:** Edged weapons cut or stab to cause serious harm, but they need more precision and skill:
    - **Kitchen Knives:** Found in every kitchen, these can be repurposed quickly. Thrust towards vital areas for maximum damage.
    - **Box Cutters:** Small and easy to hide, box cutters surprise attackers with their sharp blades. Quick slashes help create space and stop further aggression.
    - **Improvised Blades:** In desperate times, any sharp object works as an impromptu blade—think broken glass, sharpened sticks, or bits of metal—whatever's on hand to protect yourself.

MASTERING FIREARM SKILLS

Firearms are powerful tools that require adherence to basic principles and safety rules. Whether you're experienced or just starting, understanding the fundamentals of firearm training is essential for safe and effective use.

Basic Principles and Safety

- **Safety First:** Always assume a firearm is loaded. Check the chamber, magazine, and action to ensure the weapon is unloaded. Keep your finger off the trigger until you're ready to fire.
- **Know Your Firearm:** Familiarize yourself with your specific make and model. Understand all its parts and how it functions. Read the manufacturer's manual for detailed instructions.
- **Proper Stance and Grip:** Stand with your feet shoulder-width apart and slightly bend your knees for stability. Hold the firearm firmly with both hands; your dominant hand should control it while your other hand provides support.
- **Aim and Sight Alignment:** Focus on the front sight post or bead and align it with the rear sight notch or aperture. Keep it steady and centered on your target.
- **Breath Control and Trigger Squeeze:** Take a deep breath, exhale halfway, hold your breath briefly, and squeeze the trigger smoothly. Avoid jerking or flinching, as this can affect accuracy.
- **Recoil Management:** Be prepared for recoil by maintaining a firm grip throughout shooting. Brace for the kickback and quickly realign your sight on the target for follow-up shots.

Proper Handling and Shooting Skills

- **Loading and Unloading:** Always point the firearm in a safe direction when loading. Insert the magazine or load rounds into the chamber, then release the slide or bolt to chamber a round. To unload, remove the magazine, eject any rounds, and visually inspect to ensure it's empty.
- **Ready Position:** When not firing, keep the firearm in the ready position—pointed downrange or in a safe direction. Keep your fingers off the trigger but be prepared to engage if necessary.
- **Firing Drills and Practice:** Regular practice is key to improving shooting skills and building muscle memory. Start with basic drills, such as aiming, and progress to more advanced exercises like rapid fire.

- **Malfunction Clearance:** Learn how to address common issues like failure to feed or misfires. Practice clearing these malfunctions quickly and safely to minimize downtime during shooting.
- **Safe Storage and Transport:** Store firearms securely in a locked location when not in use, with ammunition stored separately. Ensure firearms are unloaded and properly secured during transport to prevent accidents or theft.
- **Continuous Education:** Stay informed about changes in gun laws, safety guidelines, and best practices by regularly consulting trusted sources such as training courses, videos, and publications.

PERSONAL PROTECTION SKILL DEVELOPMENT PROGRAMS

Local Classes & Instructors

Purpose:
Joining local self-defense classes is super valuable. You get to learn key skills in a friendly and helpful environment. When you attend classes with seasoned instructors, you pick up practical know-how and hands-on training that can be crucial in real-life situations.

Finding Local Classes:
Look for self-defense classes in your area. You might find options at martial arts dojos, community centers, or specialized self-defense schools. Use the internet, local bulletin boards, or just ask around to find good instructors and programs.

Evaluate Instructors:
Check out the instructors' qualifications, experience, and teaching style. Look for those with certifications in fields like Krav Maga, Brazilian Jiu-Jitsu (BJJ), or Taekwondo. Also, talk to current or former students to gauge how effective and professional the instructor is.

Class Structure:
These local classes usually follow a set program that helps you build skills and confidence over time. Expect a mix of theory, practical drills, and simulated scenarios to mimic real-life situations. Get ready for some tough sessions that'll push your mind and body.

Building a Strong Mindset

Purpose:
Having a strong mindset is crucial when dealing with tough situations and overcoming challenges. A strong mind helps you stay calm, focused, and ready to act when danger comes your way.

Embrace Training:
Think of training as a way to build mental toughness. Attending self-defense classes and practicing real-life scenarios can boost your confidence and resilience.

Visualize Success:
Try visualization techniques—imagine different scenarios and see yourself handling them well. This can help you think positively and feel more determined.

Stay Informed:
Keep up with current events, safety tips, and self-defense strategies through reliable sources like books or online forums. Always look for ways to learn more so you stay prepared for any threats.

Maintain Discipline:
Stick to regular training routines, stay aware of your surroundings, and always prioritize personal safety. Discipline is key for a strong mindset and it guides you both in training and real life.

PROJECT 15: INSTALL BACKYARD LIGHTS THAT DETECT MOVEMENT

Security is essential, and that is particularly important when you want to sleep. Having sensor lights can help keep your and your family safer.

Materials and Preparation

You'll want to shut off the electricity to your light fixture as you install these lights, possibly your entire house if it's not separate. You can check the wires of the lights aren't live with a circuit tester.

Before you start, you will need the following materials:

- Motion sensor lights
- Voltage Meter
- Wire connectors
- Screwdriver
- Electrical tape
- Silicone Caulk

Steps
1. Double check the wires aren't live before removing existing light fixtures.
2. Remove the lightbulbs from the previous fixture, hen unscrew the previous mount and carefully unhook the wires from the fixture, keep the screws. Keep the wires separate from each other so you can keep track of them.

Removing Older Lights

3. Connect the wires. three wires that will need to be connected to the new fixture.
   a. Connect the black fixture wire connects to the supply black wire.
   b. Connect the white fixture wire connects to the white supply wire.
   c. Connect the copper fixture wire connects to the copper supply wire.

   Apply electrical tape to connect each wire. If you're living in an older home, then you may need to use a voltage meter to identify the correct wires.

Working with the Wiring

4. Place the wires into the light fixture box, make sure none of them stick out, hen screw the fixture in, use the screws from before and a screwdriver to screw in the fixture.
5. Apply silicone caulk to seal any gaps between the wall and the box, including around the entire perimeter of the box. This will help keep the outside elements from getting in and damaging the wiring.

6. Turn the power back on to the light fixtures and test out your new motion lights. Moving in front of them should turn them on, if it doesn't then you've got some double checking to do. If you want your lights to be facing different angels, then adjust them accordingly. You may need a manual or guide for the best angels depending on where they were put.

PROJECT 16: BUILD BACKYARD TRAPS TO DETER INTRUDERS

You can keep people from getting close to your property by setting up traps around the perimeter. The simplest is modified fencing, which can be just as effective around your home as a prison. Then you can add barbed wire traps to keep people from continuing if they get passed the fencing.

Materials and Preparation

Before you start, you will need the following materials:

- Barbed wire
- No Trespassing sign
- Gloves
- Long sleeves

Steps

1. Get your roll of barbed wire and line the top of the fence with it. Be careful not to catch your hands or arms on the barbs.
2. Once you hook your wire through the chain link on your fence cut the wire and tie the ends of the wire together.
3. If you want to for extra security, add a barbed wire trench around the perimeter.
    a. Dig a 1 foot deep trench around the inside bottom of the fence line.
    b. Add barbed wire to the trench and partially burry it.
4. Add a No Trespassing sig in several places along the fence.

Barbed Wire Fencing

PROJECT 17: MAKE A TRIP WIRE ALARM FOR EXTRA SECURITY

There are a few tripwires alarms you can build. This project is to setup a doorway buzzer alarm, but you can use the steps for alarms in other locations.

Materials and Preparation

Before you start, you will need the following materials:

- Clothes pin
- Lithium battery
- Twist tie
- 2 Small pieces of cardboard
- Electric buzzer
- 2 Two-sided adhesive pads
- Fishing line
- Roll of copper tape

Steps

1. Use the twist tie to keep the clothes pin open, then wrap a piece of copper tape to the top of the opened pin.
2. Stick one of the two-side adhesive pads to the top of the copper tape on the top of the clothes pin. Then put the other one on the bottom inside of the clothes pin.
3. Put red wire to the buzzer onto the pad and make sure to firmly press down so it sticks.
4. Connect it to the buzzer and place the buzzer on top of the red wire, with the black wire pointing up.
5. Fold the black wire down into the mouth of the clothes pin and stick it onto the adhesive pad on the bottom.

Initial Setup

6. Place the lithium battery onto the black wire (negative side down) and make sure the wire and battery line up in the middle. Be sure the battery sticks to the bottom part of the clothes pin.
7. Take 1 small piece of cardboard and poke a hole in the top with a sewing needle (or pin), then put fishing wire through the cardboard and tie it. Repeat this for the second piece of cardboard.
8. Place the first cardboard into the mouth of the clothes pin and let the clothes pin gently close on top of it and the battery.
9. Take the second cardboard and tape it to a wall or banister. Somewhere that would make the trip wire difficult to see for an intruder coming through the door.
10. Test the contraption out to see if it works. Tripping on the wire should pull the cardboard out and activate the buzzer. If nothing happens, or if the string is pulled too tight, make the needed adjustments.

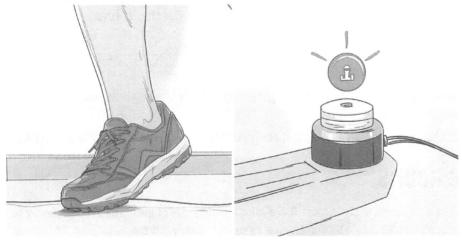

Testing the Tripwire

12. Home Security Projects

Here's how you can make your home safer and be ready for anything that comes your way.

Building super safe rooms and placing security features in smart locations can make a big difference. This chapter covers all the details on how to do that. It's really useful—you'll learn some valuable tips and tricks here.

Making safe rooms sturdy (and placing safety features smartly) is key. This part of the book dives into the nitty-gritty of protecting your home. It's important because we want you to handle whatever comes next with ease.

So, from setting up tough safe rooms to figuring out where to place your security gadgets, this chapter's got it all covered. You'll be ready for anything after reading this.

This chapter is all about making sure you know what to do. Home defense can get pretty practical, and we want you to be totally prepared.

SAFE ROOMS AND PANIC ROOMS

Imagine having a special place in your home, like a safe room or panic room. It can bring a lot of peace of mind. You and your family can feel secure during any crisis. Building such a space with care means turning a simple room into a sturdy fortress.

Purpose and Essential Features

These rooms, whether you call them safe rooms or panic rooms, are built to resist any intrusion or natural disaster.

- **Reinforced Structure:** The walls, ceiling, and door are made super strong with steel or reinforced concrete. This way, they're tough against physical hits and flying debris during storms.
- **Security Measures:** The room has robust locks, deadbolts, and security bars. This prevents anyone from getting in without permission.
- **Communication:** With two-way radios or cell phones inside, you can always reach out for help or talk to emergency services.
- **Supplies:** You should stock it with water, non-perishable food, first aid kits, and blankets. These supplies help keep everyone safe over time.
- **Ventilation:** Fresh air is important. A good ventilation system ensures constant airflow, even if you're stuck there for a while.

Location and Supplies

Where you place the safe room in your house is key to its effectiveness. Structural strength and proper placement are vital, and essential supplies matter for long-term safety.

- **Location:** The best spots are the ground floor or basement. Stay away from exterior doors and windows to reduce risk. Interior rooms or closets with few access points work great too.
- **Supplies:** Always keep enough supplies for at least 72 hours. Items like water, food, medical kits, flashlights, batteries, a portable toilet, and medications are crucial. Adding some self-defense items like a firearm or pepper spray can provide an extra layer of security.

Firearms and Other Weapons

Role in Home Defense

Purpose of Firearms:

Firearms can be essential for home defense. They help keep us safe by deterring threats or stopping them if necessary.

Immediate Threat Response:

When immediate action is required during a dangerous situation, firearms provide a powerful means of defending yourself and your loved ones.

Psychological Deterrent:

The presence of a gun can make potential intruders think twice. It's like a big sign that says, "Stay away!"

Effective Range & Lethality:

Firearms can be used from a distance, allowing you to protect your home without getting close to danger.

Training and Familiarization:

Knowing how to use firearms safely is crucial. Regular practice ensures you are prepared for any emergency.

Choosing the Right Firearm

Assessing Personal Needs:

Consider your personal needs when selecting a firearm for home defense. Factors such as your proficiency, comfort, and intended use are important.

Firearm Types:

There are various types of firearms, including handguns, shotguns, and rifles. Each has its advantages and disadvantages, and your choice should be based on what suits you best.

- **Handguns:**
  Handguns are compact and easy to handle, making them ideal for confined spaces like hallways and bedrooms.
- **Shotguns:**
  Shotguns offer strong stopping power and are effective at close range. Their spread makes them useful when precise aiming is difficult.

- **Rifles:**
  Rifles are highly accurate and perform better at long distances than handguns or shotguns. While they may be harder to use indoors, they are excellent for outdoor or distant threats.

Considerations for Ammunition:

Selecting the right ammunition is crucial. Consider factors such as caliber, bullet type, and performance when making your choice.

Alternative Weapons and Tools

Improvised Weapons:

In addition to firearms, other tools can be useful for home defense.

- **Melee Weapons:**
  Items like baseball bats, tasers, and pepper spray are effective non-lethal options for keeping intruders at bay.
- **Home Security Devices:**
  Don't overlook security gadgets. Motion sensors, reinforced doors, and cameras can help detect and deter intruders before the situation escalates.
- **Strategic Planning and Preparation:**
  Regardless of the weapons or tools you choose, planning is essential. Setting up safe rooms, securing entry points, and practicing emergency drills are key components of being fully prepared for any situation.

TACTICS FOR SECURING YOUR HOME

Being ready with a solid defense plan means you can act fast when something goes wrong. This way, you'll know what to do without wasting time figuring it out. Here are some tips to make your home safe and sound.

Nighttime Home Invasion Preparedness

Purpose: To have smart plans for keeping your home safe at night.

- **Safety Protocols:** Establish a bedtime routine. Lock all doors and windows, turn on alarm systems, and double-check everything.
- **Strategic Lighting:** Install exterior lights to deter intruders and illuminate dark areas.
- **Sleeping Arrangements:** Arrange furniture or barriers to make your bedroom safer.

Communication with Law Enforcement

Purpose: To ensure you can easily communicate with the police if something happens.

- **Emergency Contact Information:** Keep phone numbers for police and emergency services easily accessible.
- **Code Words:** Create secret words or signals to quietly alert family members or the police to danger.
- **Practice Scenarios:** Conduct practice drills with your family to prepare for emergencies.

Weapon Selection and Understanding

Purpose: To have the right tools and knowledge for self-defense.

- **Firearm Training:** Obtain professional training on how to safely handle firearms.
- **Non-Lethal Options:** Consider using pepper spray, tasers, or batons if deadly force is not necessary.
- **Legal Considerations:** Familiarize yourself with the laws regarding weapon ownership and use in your area.

Securing the Bedroom

Purpose: To create a secure and safe space in your bedroom for hiding during an invasion.
- **Reinforced Entry Points:** Strengthen bedroom doors and windows with quality locks, sturdy frames, and security bars.
- **Emergency Communication:** Keep a landline phone or panic button in the bedroom for quick access to help.
- **Safe Storage:** Store defense tools, such as firearms or sprays, safely but within easy reach.

Strategic Movement

Purpose: To move around your home swiftly and safely in case of a security breach.
- **Escape Routes:** Identify your main and backup escape paths from different areas in your home.
- **Cover and Concealment:** Locate places in your home where you can hide or take cover if there's an armed intruder.
- **Tactical Awareness:** Continuously practice being aware of your surroundings. Stay alert as you move through your home.

Verbal Commands and De-escalation

Purpose: To calm threatening situations using effective communication skills.
- **Confidence and Assertiveness:** Speak with confidence. Show you're in control when addressing potential intruders.
- **De-escalation Tactics:** Listen carefully, show empathy, and use calm, friendly language to diffuse tension.
- **Maintain Distance:** Keep a safe distance from intruders. Avoid escalating the situation unnecessarily.

Information Gathering

Purpose: To collect crucial details about potential threats and weak points in your home's defenses.
- **Security Assessment:** Regularly inspect your home's security to identify vulnerabilities.
- **Neighborhood Watch:** Collaborate with neighbors to share information about suspicious activities or concerns in the area.
- **Technology Integration:** Utilize security cameras, motion sensors, and other devices to monitor your home's surroundings.

Community Engagement and Learning

Purpose: To build a strong, resilient community against home invasion threats.
- **Neighborhood Meetings:** Organize meetings or workshops with neighbors to discuss security tips and share resources.
- **Training and Education:** Participate in self-defense classes, firearm training, and safety seminars to enhance your skills and knowledge.
- **Emergency Response Plan:** Develop a community plan for emergencies. It should outline the steps everyone will take in the event of a home invasion or other threats.

EMERGENCY COMMUNICATION AND WARNING SYSTEMS

Communication and alert systems provide warnings of potential threats and help the community come together during crises. Whether it's a break-in, natural disaster, or social unrest, having strong systems in place can boost the community's spirits.

Alarm Systems & Emergency Devices

Alarm systems are your frontline defense. They range from basic setups to sophisticated smart home integrations. Here's a breakdown:
- **Entry Sensors:** These gadgets detect unauthorized entries through doors and windows. Place them at all entry points—don't miss those ground-level windows and exterior doors! Align them correctly to minimize false alarms.
- **Motion Detectors:** These sensors spot movement inside your home. Install them in high-traffic zones like hallways and living rooms. Adjust their settings so pets or natural movements don't trigger false alerts.
- **Siren Alarms:** Loud sirens scare off intruders and alert neighbors to potential dangers. Connect them to the main alarm panel and position them in central spots for optimal sound coverage. Test them regularly!

- **Smartphone Integration:** Modern alarm systems allow you to connect via smartphone. You receive real-time alerts and can monitor your home from anywhere. Download the app and customize notifications for instant updates.
- **Backup Power:** Power outages happen, but backup batteries keep your alarm system running smoothly. Replace batteries often and perform system checks to ensure everything is functioning perfectly.

Community Collaboration

Working together makes everyone stronger! Here's how to create and maintain a collaborative spirit in your neighborhood:

- **Neighborhood Watch Programs:** Establish a watch program to get everyone involved. Hold regular meetings to discuss security concerns, share information, and plan actions.
- **Communication Channels:** Set up dedicated channels like email lists or social media groups to spread urgent news quickly. Encourage everyone to participate and share what they know.
- **Emergency Response Drills:** Regularly practice emergency drills so everyone knows where to assemble, who to call, and how to evacuate safely. Run through scenarios like break-ins, medical emergencies, and fires.
- **Resource Sharing:** Pool resources and talents to tackle common challenges and enhance preparedness. Collaborate on projects like disaster relief funds, shared emergency kits, and neighborhood cleanups.
- **Community Outreach:** Engage with local authorities, emergency services, and civic groups to strengthen the community. Participate in crime prevention programs, disaster workshops, and service projects.

READYING FOR THE UNFORESEEN

We never know what might come our way—be it a natural disaster, medical emergency, or sudden fire. But with a plan and the right supplies, we can face anything. Let's dive into how to get ready for fire safety, emergency escape plans, and first aid.

Fire Safety and Prevention

Fires are scary, but guess what? With some simple steps, we can prevent and manage them.

- **Install Smoke Detectors:** These handy gadgets are our first line of defense against fires. Place smoke detectors everywhere you can think of (especially near bedrooms).
- **Keep Fire Extinguishers Handy:** Every home should have at least one easy-to-reach fire extinguisher. Show your family how to use them—it's super important!
- **Adopt Safe Cooking Practices:** The kitchen is where most fires start. Always be careful when cooking. Don't leave the stove unattended, and keep flammable items away from heat.
- **Maintain Electrical Systems:** Overloading circuits or having bad wiring can start fires too. Check and maintain your electrical systems regularly to avoid those risks.

Emergency Escape Plans

A good escape plan works wonders in many situations.

- **Identify Escape Routes:** Spend time figuring out multiple ways to get out of every room in your home. Make sure everyone knows these routes well.
- **Practice Drills Regularly:** Have fire drills with your family often. This way, everyone knows what to do if something happens. Practice different scenarios too, like meeting up at a specific spot outside.
- **Designate Meeting Points:** Decide on a safe spot outside where everyone will meet after escaping. This helps you see who's safe and who might still need help.
- **Communicate with Emergency Services:** Teach everyone how to call emergency services (like 911 in the US). Make sure they know what information to give the operator.

First Aid and Medical Supplies

Being ready with first aid supplies is crucial during emergencies.

- **Create a First Aid Kit:** Gather all the basics in a first aid kit—things like bandages, antiseptic wipes, scissors, tweezers, and painkillers. Don't forget any special medications your family might need.

- **Learn Basic First Aid Skills:** It's always good to know some first aid skills—like CPR, wound care, and treating burns or fractures. You can find online resources or local courses to learn these skills.
- **Regularly Check and Restock Supplies:** Keep your first aid kit up-to-date and fully stocked. Check expiration dates and replace items that are used or expired right away.
- **Consider Specialized Medical Needs:** If someone in your household has unique medical needs (like allergies or chronic conditions), make sure your first aid kit has the necessary medications and supplies for them too.

PROJECT 18: DIY HIDDEN GUN STORAGE SHELF

During a crisis, it can be useful to have a gun more easily accessible, but not obvious. Adding a hidden gun storage shelf can give you some peace of mind because only you will know that there is a gun present that you can use to project yourself and your family.

Materials and Preparation

Before you start, you will need the following materials:

- 1 x 8 x 8 in wood board
- 1 x 6 x 6 in wood board
- 2 x4 x in wood board
- 2 overlay surface mount cabinet hinges
- Polyurethane
- 1.5 in trim head screws
- Wood stain
- Wood filler
- Ear protection
- Drill
- Masks
- Safety goggles
- Saw
- Sanding block or sandpaper
- Tape measure

Steps

1. Cut the wood into pieces:
    a. Cut the 1 x 8 into 2 pieces that are 47.75 in long
    b. Cut the 1 x 6 into three pieces:
        i. 1 piece 49.25 in long
        ii. 2 pieces 7.25 in long
    c. Cut the 2 x 4 so it is 47.5 in long

The Cut Wood Pieces

2. Lay the 1 x 8 x 47.75 in piece and both 1 x 6 x 7.25 in pieces on their sides with the small pieces on either end of the long piece.
3. Drill two holes in both of the small pieces. Make the holes about 1 in from the top (where the long piece will attach). Drill through the long piece so that the holes align.
4. Screw in the 1.5 in screws in all four newly drilled holes.
5. Lay the long board down flat.
6. Place the remaining 1 x 6 x 49.25 in piece along the top of the board you just lay down.
7. Close the hinges and place them on either side of the loose board.

Add the Hinges

8. Secure both hinges in place.

9. Secure the remaining 1 x 8 x 47.75 in piece to the top.
10. Repeat steps 3 and 4 to secure it in place.
11. Test the door with the hinges work by opening it.
12. Add the wood filler over the holes.
13. Sand any rough edges. Use the mask so that you don't get wood dust in your lungs.
14. Add wood stain to the final product if you want.

When you finish, you can place the shelf anywhere, whether on the wall or on top of furniture.

Finished Secret Shelf

PROJECT 19: HOW TO BUILD YOUR OWN FARADAY CAGE

The Faraday cage can be used to block some times of electromagnetic fields because it disrupts the signals. . If you don't want to buy one, you can easily make your own.

Materials and Preparation

Before you start, you will need the following materials:

- Metal mesh screen
- 5 8 in wood strips
- Alligator slip cable
- Staple gun
- Hack saw
- Ruler and heavy duty scissors

Steps

1. Cut out a 8 x 16 in rectangle from the metal mesh screen.
2. Lay the 8 x 16 in metal mesh rolled out on a flat surface.
3. Staple the metal mesh to the wood strips.
    a. Staple one strip to the end of the mesh.
    b. Staple the second strip about 5.5 in away from the first strip.
    c. Staple the third strip about 2.5 in away from the second strip.

d. Staple the fourth strip about 5.5 in away from the third strip
   e. Staple the last strip at the end of the mesh.
4. Fold the mesh at each of the metal strips to form the Faraday cage.

The Finished Cage

13. Emergency Communication Techniques

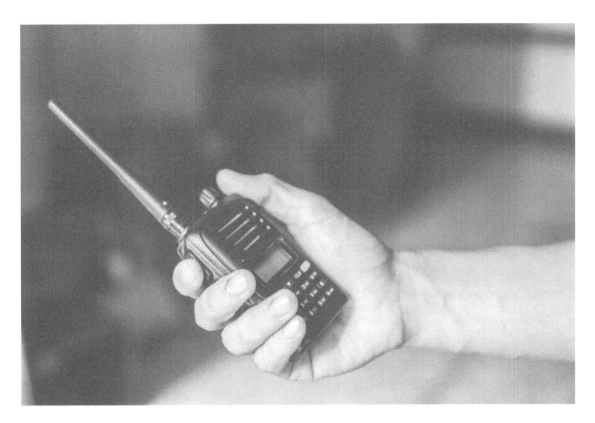

SIGNIFICANCE OF COLLECTING AND SHARING INFORMATION DURING CRISIS SITUATIONS

When it comes to surviving tough times, knowing how to gather and share information is super important. Here's why:

- **Real-Time Updates:** Staying in the loop with what's going on around you—and even far away—is key for making smart choices. Whether it's a big storm or something else, having the latest info means you can react and adapt quickly.
- **Safety Alerts:** Emergency alerts give you critical info like where to go for shelter or when to evacuate. Staying connected ensures you know about any dangers nearby so you can keep yourself and your loved ones safe.
- **Coordination with Others:** During a crisis, communication with others becomes crucial. You'll need to coordinate with emergency workers and folks in your community. By sharing what you know and pooling resources, everyone has a better shot at getting through it.
- **Maintaining Calm:** Information can also help keep everyone calm. When people know what's happening and what steps to take, they're less likely to panic. This means fewer rash decisions that could make things worse.

Objectives of Emergency Communication

So, what are we aiming for with emergency communication? Here's the lowdown:

- **Receive Information:** You need to get updates on what's going on—whether it's the weather, evacuation orders, or safety notices. This is essential for making good decisions and knowing what you should do next.
- **Local Communication:** Set up ways to talk with people nearby—family, neighbors, or community groups. Sharing info and resources locally can be a lifesaver during an emergency.
- **External Communication:** It's also important to keep lines open beyond your local area. Maybe you need help from authorities or want to share info outside your immediate circle. This boosts resilience and helps ensure survival.

ESSENTIAL COMMUNICATION METHODS

Cell Phone

- **Purpose:** The cell phone is a handy tool for communication during emergencies. You can call, text, and check the internet for information.
- **Usage:**
    - To call someone, unlock your phone. Then, tap the Phone app. Enter the number and hit call.
    - To send a text, open the Messages app. Choose who you want to text, type your message, and tap send.
    - To surf the web, tap the browser icon. Type the website URL in the address bar and press Enter.
- **Tip:** If there's a network jam, try texting instead of calling. Texts use less bandwidth.

Signaling Mirrors

- **Purpose:** These mirrors are super helpful for getting attention from far away, especially in remote areas where other communication methods might not work.
- **Usage:**
    - Hold your mirror up to the sun and aim it at something like a passing airplane or a search team.
    - Flash it three short times followed by a longer pause to create a clear signal pattern.
- **Tip:** Practice using your signaling mirror in different lighting conditions so you know how well it works.

Whistles

- **Purpose:** Whistles are great for calling for help when it's hard to see or shout.
- **Usage:**
    - Blow the whistle in short, loud bursts to get noticed.
    - Use three short blasts followed by a pause to signal that you need help.
- **Tip:** Clip a whistle to your backpack or clothing so it's easily accessible in an emergency.

Walkie-Talkies

- **Purpose:** Walkie-talkies are perfect for staying in touch with your group. They help coordinate tasks and maintain communication during emergencies.
- **Usage:**
    - Press the power button to turn on the walkie-talkie.
    - Select a channel using the selector knob or buttons.
    - Push and hold the talk button when speaking; release it when listening.
    - Keep your messages short and clear to avoid confusion.
- **Tip:** Assign different channels for different groups or individuals to keep communication organized.

ENHANCED COMMUNICATION SOLUTIONS

Ham Radio

Amateur radios, or ham radios, operate on special radio frequencies allocated by the government for non-commercial use. Here's how you can make the most of a ham radio:

- **Obtain a Ham Radio License:** Before using a ham radio, you must get a license. This ensures you understand the rules and regulations for legal operation.
- **Selecting a Radio:** There are many types of ham radios, ranging from handheld devices to base stations. For beginners, a handheld transceiver (HT) like the Baofeng UV-5R is popular. It's affordable, versatile, and perfect for starters!
- **Learning Basic Operation:** Get to know your ham radio. Learn how to tune frequencies, adjust power levels, and program channels. Practice sending and receiving messages to build confidence.
- **Connecting with Local Hams:** Join online forums or local clubs to get advice from seasoned operators. Connecting with other hams enhances your capacity for emergency communication.
- **Emergency Preparedness Programming:** Program your radio with emergency frequencies and repeater channels. This helps you access crucial information and coordinate during crises. Ensure your radio has NOAA Weather Radio channels for weather updates.

Hand-Crank Emergency Radios

Hand-crank emergency radios are essential for staying updated during power outages and emergencies. These radios typically offer multiple power sources—hand-crank charging, solar panels, and battery backup.

Here's how to make the best use of them:

- **Understanding Functionality:** Learn about the features of your hand-crank emergency radio, such as AM/FM reception, NOAA Weather Radio capabilities, flashlight, and USB ports.
- **Hand-Crank Charging:** No electricity? No problem! Use the hand crank to generate power by rotating it steadily to charge the internal battery.
- **Solar Panel Charging:** Use the built-in solar panels to gather energy from sunlight. Place the radio in direct sunlight to recharge the battery—eco-friendly and sustainable!
- **Battery Backup:** Keep extra batteries handy to support hand-crank and solar charging methods. Store them fully charged in a cool place to maintain effectiveness during emergencies.
- **Programming Channels:** Set essential frequencies and channels on your emergency radio—local emergency services, NOAA Weather Radio stations, and amateur repeaters. Update these regularly to stay informed about changing situations.

Satellite Communication Devices

Satellite communication devices provide reliable connectivity in remote areas or during large-scale disasters when regular networks fail.

Here's how to use satellite communication devices effectively:

- **Selecting a Device:** Choose wisely after considering coverage area, data speeds, and subscription plans. Garmin and SPOT are well-known brands.
- **Subscription Plans:** Pick a subscription plan based on your needs. These plans ensure you can access necessary services with different data allowances and messaging capabilities.
- **Messaging Features:** Get familiar with messaging features like text messaging, email, and SOS functionality on your device. Practice sending and receiving messages to communicate effectively during emergencies.
- **GPS Tracking:** Use the built-in GPS tracking feature to share your location with trusted contacts or responders. This aids in efficient coordination and enhances safety during outdoor activities or emergencies.

CRISIS COMMUNICATION PLANS FOR PREPAREDNESS

Being prepared means having reliable ways to stay in touch with loved ones and emergency services. It's vital. So, let's go over key communication tools and protocols to keep connected during disasters.

Protocols for Maintaining Contact

Establish a Communication Plan:

Creating a solid communication plan is crucial. Outline contact steps and set meeting spots if you get separated. Make sure everyone in the family or group knows the plan inside out.

Backup Contacts:

Store important numbers, not just on your phone but also on paper in your emergency kit. Include family members, a designated out-of-area contact, and local emergency agencies.

Practice Regularly:

Keep practicing your communication methods. Hold regular drills with your family or group to ensure everyone knows what to do when it counts.

Power Source Planning

Solar Panels:

Use the sun! Portable solar panels can keep your devices charged. Place them in direct sunlight to charge power banks or directly power devices.

Portable Solar Generators:

For more extended power needs, think about investing in a portable solar generator. These handy devices provide reliable power for multiple gadgets all at once. Pick one with enough capacity for all your communication needs.

DEVELOPING EMERGENCY COMMUNICATION STRATEGIES

Scenario-Based Protocols:

- **Think about Different Scenarios:** Imagine all sorts of emergencies like natural disasters, power outages, or even social chaos. Adjust your communication plans for each case.
- **Set Meeting Spots:** Decide on specific places where family or group members can gather if you get separated.
- **Communication Order:** Make a list of who to call first during emergencies. This ensures information spreads quickly and efficiently.
- **Emergency Contact List:** Gather a list of key contacts—local authorities, family, neighbors—and keep it in easy-to-find spots.

Testing and Practice:

- **Regular Drills:** Conduct frequent drills so everyone becomes familiar with the protocols and gear.
- **Feedback Helps:** Ask for feedback after each drill to identify areas for improvement. Then, fine-tune your plans accordingly.
- **Kit Check-Ups:** Regularly check your emergency kits to ensure everything works and is fully charged.

FACTORS TO SAFEGUARD COMMUNICATION EQUIPMENT AGAINST DAMAGE OR INTERFERENCE

Keeping your communication equipment safe is super important, especially during emergencies. Make sure you follow these key tips to keep everything in top shape:

Secure Storage:

First, make sure to store your communication devices in a secure, waterproof container. This keeps them safe from water, impact, and debris. Use a sturdy case that's also shockproof. Trust me, this will help a lot during transport or storage. You'll be glad you did.

Electromagnetic Shielding:

Next up, shield your devices from electromagnetic interference. Why? Because it messes with the signals. Use metal enclosures or special cases. This is critical for keeping your gear working properly.

Backup Power:

Also, make sure your devices have a reliable power source. Rechargeable batteries and portable solar panels are the way to go. In emergencies, power can be scarce, so having backup power is essential for staying connected.

Signal Redundancy:

Don't rely on just one method of communication. Mix it up! Use two-way radios and modern satellite devices together. This way, if one fails, the other can still keep you connected.

Regular Maintenance:

Lastly, regularly check and maintain your equipment. Inspect them often to ensure they're working fine. Clean contacts, replace old batteries, and update software when needed. Doing this keeps everything running smoothly.

14. Fundamental Knots for Every Situation

THE SQUARE KNOT

The Square Knot (sometimes called the Reef Knot) is essential. It's versatile and super easy to tie. Let's see how to tie it and why preppers should care:

Tying the Square Knot:

1. First, grab two ends of a rope or cord.
2. Take the right end and cross it over the left, making an overhand knot.
3. Then, cross the left end over the right one.
4. Pull both ends tightly to secure your knot.

Importance to Preppers:

- **Versatile Binding:** When you need to secure items or build makeshift structures, the Square Knot comes to the rescue. Whether constructing a shelter, securing gear, or bundling supplies, this knot is quick and reliable.
- **Emergency Repairs:** In a pinch? Use this knot for quick fixes! Repair torn clothing, mend straps, or even secure a splint during medical emergencies.
- **Medical Applications:** The Square Knot isn't just for ropes—it's handy in first aid too! Secure bandages or slings to provide stability and support for injured limbs.

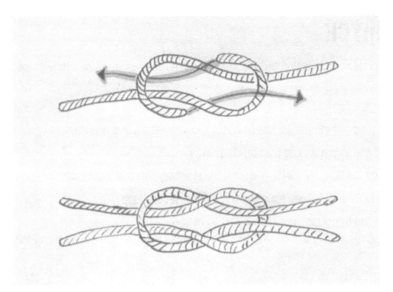

The Square Knot is a simple yet highly effective tool. It's perfect for all sorts of binding and securing needs. In times of need, this knot helps you stay prepared and adaptable.

THE CLOVE HITCH

It's a super handy knot with lots of uses! Here's how to tie it and why it's crucial for prepping:

Tying the Clove Hitch:

1. Start by draping the rope over whatever you need to secure. Easy, right?
2. Make an "X" shape by crossing the rope over itself.
3. (Almost done!) Take the end of the rope, slide it under the object, and put it through the loop made by that "X".
4. Now pull both ends of the rope in opposite directions—tighten that hitch!

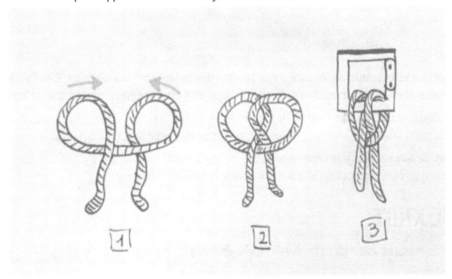

Importance to Preppers:

- **Securing Gear:** Need to quickly tie down tarps? Or secure gear to your car or anchor ropes to trees & poles? The Clove Hitch is your go-to!
- **Versatility:** This knot can be tied in the **middle** of a rope. Yes, really! That means you can create adjustable points for rigging shelters, makeshift clotheslines, or hanging stuff to dry or store.
- **Emergency Repairs:** Got broken equipment? Need an urgent fix? The Clove Hitch is perfect for speedy repairs. It's great for temporarily securing things or creating supports (or even structures).

THE TAUT LINE HITCH

The Taut Line Hitch is amazing for adjusting tension and securing lines in all sorts of situations. Here's how you can tie it and why it's crucial for preppers:

Tying the Taut Line Hitch:

1. First, wrap the rope around something. It could be a tent peg or maybe a tree branch.
2. Next, take the free end and loop it around the standing part.
3. Bring the free end inside the loop, then wrap it around the standing part one more time.
4. Then, slide the free end through the loop once again, making a second loop.
5. Finally, pull on the free end to tighten your knot. Adjust as needed.

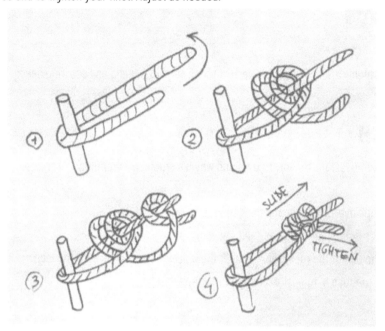

Importance to Preppers:

- **Adjustable Tension:** When setting up shelters, traps, or other things, this hitch is a lifesaver. With the Taut Line Hitch, you can easily tweak tension to keep everything stable. Whether dealing with wind or other elements, your setup stays strong.
- **Versatility:** Setting up a tent? Hanging a tarp? Or maybe creating a clothesline? This knot does it all. It adapts to your needs like magic. In any situation, whether camping or facing an emergency, this hitch will come in handy.
- **Ease of Untying:** Secure yet simple to untie—what more could you want? You can quickly adjust or take down your setup whenever you need to. This is super helpful when you have to move fast in an emergency.

THE BOWLINE KNOT

The Bowline Knot is reliable, versatile, and easy to tie. Here's how to tie it and its importance for prepping:

Tying the Bowline Knot:

1. First, make a small loop in the rope. Keep a long tail for the working end.
2. Next, pass the working end through the loop from underneath. This forms a crossing point.
3. Wrap the working end around the standing part of the rope.
4. Then, feed the working end back down through the loop.
5. Pull both the standing part and the working end at once to tighten it.
6. Make sure there's a secure loop with the working end exiting from the bottom.

Importance to Preppers:

- **Secure Attachment Point:** The Bowline makes a fixed loop at the rope's end—perfect for a secure attachment point. Use it to secure shelters, create harnesses, or lower and hoist equipment.
- **Reliability:** Unlike other knots, the Bowline keeps its strength under tension. It's great for critical tasks like rescue missions or securing heavy loads.
- **Ease of Untying:** Despite being strong, it's still easy to untie—even after bearing a load. That's handy when you need to quickly adjust or dismantle without fighting annoying knots.
- **Emergency Rescue:** Think wilderness survival or rescue missions. The Bowline helps create lifelines, secure during rappelling, or make stretchers for injured folks.

THE SHEET BEND KNOT

The Sheet Bend Knot is a solid way to join two ropes of different sizes. Here's how you tie it and why it's important for preppers:

Tying the Sheet Bend:

1. First, take the thicker rope and make a loop (or "bight").
2. Next, pass the thinner rope through the bight from below.
3. Then, wrap the thinner rope around the back of the bight and tuck it under itself.
4. Finally, pull both ends to tighten the knot securely.

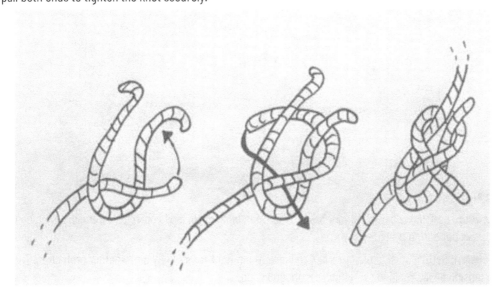

Importance to Preppers:

Adaptable Joining:

The Sheet Bend is adaptable and secure for joining ropes of various sizes. This allows you to create longer ropes or build stronger structures.

Emergency Rigging:

When setting up shelters or making tools quickly, you might need to rig materials fast using what's on hand. The Sheet Bend helps you join ropes swiftly to construct key structures for shelter, transport, or rescue missions.

Resilient Connections:

The Sheet Bend forms a strong and reliable link between ropes, ensuring durability in tough conditions. Whether you're securing gear on a trip or building essential structures in survival situations, this knot will help create connections that can handle stress and pressure excellently.

THE TRUCKER'S HITCH

The Trucker's Hitch is a must-know knot. It creates a solid and adjustable tensioning system. Let's dive into how to tie it and why preppers need it.

Tying the Trucker's Hitch:

1. **Make a Loop:** Start by making a loop in the rope near what you're securing.
2. **Anchor the Rope:** Pass the rope's free end through or around your anchor point, like a tree or post, then back through your loop to make a slip knot. Tighten the slip knot hard against the anchor point.
3. **Form a Bight:** Next, make another loop in the rope's free end, forming a bight.
4. **Create a Pulley System:** Push this bight through the first loop from step one, making a new loop. Pull the bight through that initial loop, so it acts like a pulley.
5. **Adjust Tension:** Finally, pull on the rope's free end to get the tension you want. Adjust as needed for the perfect tightness.

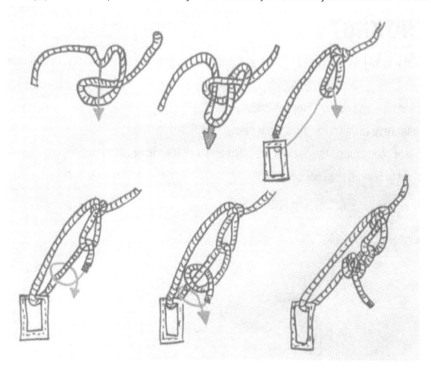

Importance to Preppers:

- **Securing Loads:** The Trucker's Hitch is your best friend when tying down loads on vehicles, trailers, or even makeshift carts. Your cargo stays put during transit—no worries.
- **Creating Tension:** Setting up shelters or tarps? You need them tight for stability and weather protection. This knot lets you adjust rope tension easily, making everything secure and snug.

- **Versatility:** Beyond tying down loads and shelters, this knot does more. Build tripwire alarms, makeshift clotheslines, or rig up improvised tools. The possibilities are endless with the Trucker's Hitch in your toolkit.

FIGURE EIGHT KNOT (STOPPER KNOT)

The Figure Eight Knot, sometimes called the Stopper Knot, is a key knot known for its flexibility and trustworthiness. Here's how to tie it and why it's crucial for prepping.

Tying the Figure Eight Knot:

1. First, make a loop in the rope.
2. Then, take the working end around and push it through the loop, forming a simple overhand knot.
3. Next, guide the working end back through the loop, mirroring the path of that initial knot.
4. Finally, tighten by pulling both ends of the rope.

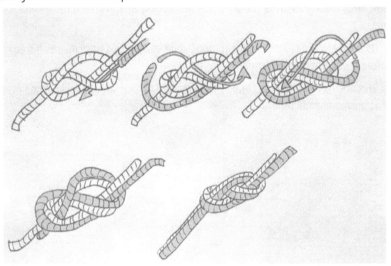

Importance to Preppers:

- **Securing Equipment:** The Figure Eight Knot shines as a stopper knot—it keeps your rope locked in place. This prevents slippage through grommets, loops, or anchor points.
- **Creating Loops:** You can use this knot to make large loops for various objects or small loops, like with carabiners. It's great for anchor points or forming adjustable lines; it serves as a reliable attachment point.
- **Emergency Repairs:** In emergencies, this knot is super handy. Use it to isolate damaged rope sections by tying stopper knots above and below the damage. That way, you keep the rest of the rope intact.

THE PRUSIK KNOT

The Prusik Knot is perfect for making adjustable loops and securing ropes in all kinds of scenarios. Here's a quick guide on tying it and why it's crucial for preppers:

Tying the Prusik Knot:

1. Start with a piece of cord or rope, called the "Prusik loop."
2. Make a small loop by overlapping the two ends of the loop.
3. Pass this loop around the main rope or cord you want to attach to, keeping the loop parallel to that main rope.
4. Wrap the free ends of the loop around the main rope several times. (How many? It depends on how much friction and security you need.)
5. Thread those ends back through the small loop you made in step 2.
6. Tighten it up by pulling both ends of the loop and adjusting the wraps as needed until it grips just right.

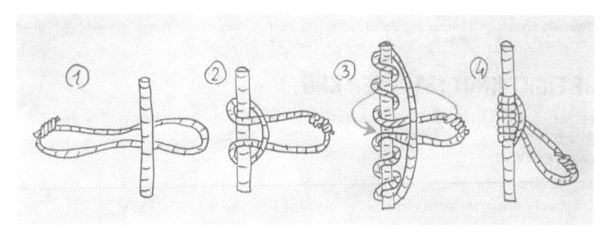

Importance to Preppers:

- **Versatile Attachment:** The Prusik knot gives you a flexible and strong way to attach things, making adjustable loops that hold tight to your main rope.

- **Adjustable Tension:** Its ability to slide along the main rope (while still holding firm) means it's easy and secure to adjust tension—great for different conditions and terrains.

- **Emergency Rescue:** This knot is essential in climbing systems and rescue situations, allowing preppers to create solid safety backups and controlled movements on ropes.

15. Survival Gardening

WHY GROW YOUR OWN FOOD?

In uncertain times, growing your own food is a key skill for self-sufficiency and resilience. Survival gardening ensures you have a reliable food source even when things go wrong.

By cultivating fruits, veggies, and herbs, you can protect yourself from food shortages, economic problems, and supply chain issues. Learn how to grow nutrient-rich food right in your backyard, balcony—or even indoors.

Take control of your food security. Reconnect with the earth. Start your journey of self-reliance with survival gardening today.

EVALUATING YOUR GARDEN AREA

Evaluating Available Space for Gardening:

Step 1: Survey Your Outdoor Space

Imagine rows of veggies and lush greenery. Walk around your yard and look for areas with lots of sunlight and those shaded by trees or buildings.

Step 2: Sunlight Exposure:

Check how much sunlight each spot gets daily. Watch the sun's path for a few days, noting where it shines brightest and the duration. Ideal garden spots need sunlight for six to eight hours each day to boost plant growth.

Step 3: Airflow Considerations:

Notice how air moves in your yard. Good airflow helps keep plant diseases away. See if anything, like fences or houses, blocks the wind. Aim for a mix of protected spots and open spaces to help your plants grow well.

Identifying Potential Soil Challenges:

- **Soil Quality Assessment**
- **Testing for Compacted Soil**
- **Addressing Soil Challenges**

Let's dive into these details so you can get the best results in your garden.

OPTIMIZING SUNLIGHT AND WIND UTILIZATION EFFICIENCY

Positioning Garden Beds for Maximum Sunlight Exposure:

1. **Check Your Garden Space:**
   First, evaluate your garden area to spot the best places with optimal sunlight. Look for spots that get at least 6 to 8 hours of direct sunlight each day.

2. **South-facing Orientation:**
   Aim your garden beds south to grab the most sunlight all day long. South-facing gardens soak up sunlight from morning till evening, giving plants perfect growing conditions.

3. **Avoid Shading from Structures:**
   Watch out for structures like houses, garages, fences, and trees that can cast shadows over your garden. Trim back branches or move garden beds to sunnier spots if needed.

4. **Use Trellises for Vertical Gardening:**
   Trellises let you grow vines vertically, using space better and boosting sunlight exposure. Train plants like tomatoes, cucumbers, and beans to climb, freeing up ground space for other crops.

5. **Strategic Row Orientation:**
   Plant rows perpendicular to the sun's path so every plant gets enough light. This keeps taller plants from shading shorter ones, optimizing growth and productivity.

Promoting Airflow for Disease Prevention:

1. **Importance of Airflow:**
   Adequate airflow stops moisture and humidity buildup, which leads to diseases and pests. Good airflow helps plants breathe, reducing stress and disease risk.

2. **Spacing Between Plants:**
   Keep enough space between plants for good airflow around each one. Crowded plants fight for sunlight and block airflow, creating disease-friendly conditions. Ensure there's enough space to avoid these problems as much as possible.

3. **Strategic Placement of Garden Beds:**
   Don't place garden beds too close to structures or thick vegetation; this blocks airflow. Leave gaps between beds and nearby structures for natural air circulation.

4. **Add Open Areas:**
   Design your garden with open areas or paths between beds to help airflow through the garden. These spaces act as air channels, dissipating extra moisture and heat.

5. **Use Windbreaks:**
   While airflow is key, too much wind can harm plants. Place windbreaks like fences, shrubs, or trellises strategically to protect delicate plants from strong winds without blocking overall airflow in the garden.

HANDLING FIXED STRUCTURES

Managing Shading from Immovable Structures

- **Assessing Shade Patterns:** Take a peek at how the sunlight dances around your garden. Notice where shadows linger, thanks to nearby structures like houses, garages, or fences.
- **Identifying Shade Sources:** Pinpoint the main culprits of shading and see how they affect different parts of your garden. Houses and garages usually cast shadows at certain times of the day, while fences might throw shade more locally.
- **Trimming Trees & Bushes:** If trees or bushes are contributing to the shading, consider trimming them. Focus on branches that block sunlight during peak growing hours.
- **Strategic Plant Placement:** Place taller plants or trellises wisely to counteract shading from those stubborn structures. Position taller crops on the side opposite the shadows so they can soak up the most sun.

Techniques for Optimizing Plant Growth

- **Vertical Gardening Solutions:** Try vertical gardening with trellises, arbors, or stakes for climbing plants. This allows plants to reach sunlight without being obstructed by nearby structures.
- **Raised Beds & Containers:** Use raised beds or containers in shaded spots to create cozy microclimates for sun-loving plants. Move these around to catch sunlight from different angles throughout the day.
- **Selection of Shade-Tolerant Varieties:** Choose plants that thrive in partial shade or dappled sunlight. Leafy greens, herbs, and some root vegetables often do well with less sun and still produce great harvests.
- **Implementing Reflective Surfaces:** Install reflective fences or walls—like white or light-colored ones—to bounce sunlight back into shadowed areas of your garden. This simple trick can boost sunlight exposure and improve plant growth.

Additional Tips for Managing Immovable Structures

- **Observation & Adjustment:** Keep an eye on how shade patterns change and adjust as needed. As seasons shift and sun angles change, some spots might receive more or less shade, so be ready to adapt.
- **Utilizing Shade-Loving Plants:** Embrace the unique conditions created by immovable structures by growing shade-loving plants in these areas. Ferns, hostas, and certain ornamental grasses thrive in shady spots and can enhance the beauty of your garden.
- **Maximizing Sun Exposure:** Make the most of open spaces that get plenty of sun all day long. Reserve these sunny spots for crops that need full sun to thrive.
- **Strategic Planting Calendar:** Plan your planting schedule based on changing sunlight patterns. Start with cool-season crops in shaded areas during early spring, and switch to warm-season crops as summer brings stronger sunlight.

MANAGING SOIL AND FERTILITY

Understanding Soil Compaction

When soil gets too compacted, it stops air, water, and nutrients from moving around easily, making it hard for plants to grow well. Roots can't develop properly either, so the plant struggles to get the food and water it needs from the soil. To see if your soil is compacted, try the wire test: just take a wire or even a coat hanger and push it into the ground. If it hits resistance after just a few inches, your soil might be compacted and needs some attention.

Strategies for Improving Soil Quality

Composting:

Organic waste turns into compost (a super-rich soil booster) through composting. Start by collecting kitchen scraps like eggshells, coffee grounds, and fruit & veggie peels. Layer these organic materials in a compost bin or pile, alternating between green stuff (high in nitrogen) and brown stuff (high in carbon), such as leaves & straw. Keep turning the compost so it stays moist and gets air. In a few months, you'll have lovely dark compost that will greatly improve your garden soil.

Mulching:

Mulching is easy but really helps soil quality and fertility. Plus, it saves water and stops weeds! Put down a layer of organic mulch like shredded leaves, straw, or grass clippings around your plants. This mulch will slowly break down, adding organic matter to the soil and attracting beneficial critters like earthworms. Mulch also helps control soil temperature—keeping it cooler in summer and warmer in winter—making plants grow better.

Proper Fertilization:

Fertilizing helps by adding nutrients back into the soil for strong plant growth. First, do a soil test to check pH levels and see what nutrients are missing before you add any fertilizers. Choose a balanced fertilizer with the right amounts of potassium (K), phosphorus (P), and nitrogen (N) for your plants. Follow the recommended rates and timing carefully when applying fertilizer. Don't overdo it—it can harm plants and cause nutrients to spread too far out.

Consider using organic fertilizers like compost tea, fish emulsion, or seaweed extract to help improve soil structure over time and provide slow-release nutrients.

CHOOSING PLANTS AND PLANNING CROPS

Selecting Appropriate Crops for Survival Gardening

- **Understanding Climate and Soil Conditions:** Do some homework on your area's climate and soil conditions to figure out which crops will do well. Consider factors like soil pH, rainfall, and typical temperatures.
- **Choosing Resilient Varieties:** Pick crops that are well-suited to your climate and soil type. Look for resilient varieties that can handle environmental stresses and resist diseases.
- **Prioritizing High-Yield Crops:** Opt for high-yield crops that produce a lot of food per square foot, such as tomatoes, beans, squash, and leafy greens (like spinach and kale).
- **Selecting Space-Efficient Plants:** Since space is a consideration, choose plants that don't require much room and can grow vertically or in small spaces. Vining crops like cucumbers and peas are excellent for vertical gardening.
- **Considering Nutritional Value:** Include a variety of crops to ensure a range of essential nutrients. Incorporate fruits, vegetables, grains, and legumes for a balanced diet.

Planning Crop Rotations and Companion Planting

- **Creating a Crop Rotation Plan:** Develop a crop rotation plan based on your crops' needs and the layout of your garden. Rotate crops according to their nutrient requirements and pest issues.
- **Incorporating Cover Crops:** Consider adding cover crops like clover or rye between your main crops. They help maintain soil health, suppress weeds, and prevent erosion. Additionally, they add organic matter to the soil when tilled under.
- **Utilizing Companion Planting:** Companion planting involves growing certain plants together to enhance their growth and deter pests. For example, planting basil near tomatoes can improve their flavor and repel pests, while marigolds can help chase away nematodes.
- **Choosing Complementary Plant Combinations:** Select plant pairs that complement each other's nutrient needs and growth patterns. For instance, planting beans (which fix nitrogen) next to corn (a heavy feeder) can boost soil fertility and overall plant health.
- **Avoiding Planting Similar Crops Together:** To prevent the buildup of pests and diseases, avoid planting the same type of crops in the same spot every season. Rotate them to different parts of the garden each year.

Monitoring and Adjusting:

Keep a close eye on your garden. Observe how different crops interact with each other and their environment. Be prepared to adjust your crop rotation and companion planting strategies based on your observations and the needs of your plants.

CONTROL OF PESTS AND PREVENTION OF DISEASES

Integrated Pest Management

Purpose: Integrated Pest Management (IPM) is an approach that focuses on managing pests without relying on harmful chemicals. By using different techniques, we can control pests while protecting helpful insects and the environment.

Identify Pests: First, find out what insects are bugging your garden. Watch for common pests like aphids, caterpillars, and beetles. Keep an eye on your plants regularly so you can catch pest problems early.

Beneficial Insects: Encourage good bugs! Parasitic wasps, lacewings, ladybugs, and other friendly insects help control pest numbers.

Crop Rotation: Change up your crops once a year to disrupt pest life cycles and reduce the number of pests in your soil. By planting different crops in the same spot each year, you prevent pests from settling in.

Companion Planting: Grow certain plants together to keep pests away or attract good bugs. For example, planting marigolds next to veggies helps keep nasty nematodes and other pests at bay.

Physical Barriers: Use barriers like row covers or netting to protect plants from birds and insects. These barriers act like shields, keeping pests away from your crops.

Sanitation Practices

Purpose: Good sanitation keeps plant diseases from spreading by removing sick plant material and reducing opportunities for germs to thrive.

Clean Tools: Clean gardening tools often to stop disease from spreading. Use a mix of bleach and water to disinfect used tools.

Remove Diseased Plants: Quickly remove any sick plants in your garden to prevent them from spreading germs. Dispose of diseased plants properly, either by throwing them away or burning them to avoid further contamination.

Prune Infected Branches: Cut off infected branches or leaves to prevent disease spread. Use sharp pruning shears for clean cuts that minimize plant damage.

Practice Good Hygiene: Wash your hands well after handling sick plants to avoid spreading germs to healthy ones. Also, avoid working in the garden when plants are wet, as this can spread fungal diseases.

Trap Gardening Techniques

Purpose: Trap gardening involves planting specific crops that attract pests away from your valuable plants, allowing you to control their numbers without using chemical pesticides.

Trap Crops: Plant trap crops like radishes or mustard greens around your garden's edge to lure pests away from main crops. Pests are drawn to these trap crops instead of your important plants.

Monitoring: Check trap crops often for signs of pest activity—such as chewed leaves or other damage clues. This helps you assess the severity of the pest problem and decide on the next steps.

Control Methods: Once pests are on the trap crops, use non-chemical methods to get rid of them. Options include hand-picking bugs, using insecticidal soap, or relying on predators like predatory insects.

Crop Rotation: Rotate trap crops yearly so pests don't get accustomed to them. Planting different trap crops each year helps keep pest numbers down and reduces the risk of resistance development.

PRESERVING AND STORING SEEDS

Seed saving—it's something that can keep your food growing strong and give you seed sovereignty. This means less need (or none at all) for buying seeds from others, allowing you to have a garden that sustains itself. Let's dive into why seed saving matters and how to collect, dry, and store seeds to keep your food secure.

Importance of Seed Saving

Saving seeds from your plants keeps different varieties alive and helps your food stay resilient. Here's why it's key:

- **Long-Term Sustainability:** By saving seeds from your best plants, you preserve traits like disease resistance, high yields, and delicious flavors, ensuring a bountiful garden.
- **Seed Sovereignty:** When you save seeds yourself, you're in control of your food supply and don't have to rely on purchasing seeds. You can grow plants that thrive in your local climate and suit your tastes.

Techniques for Seed Saving

Let's explore how to collect, dry, and store seeds effectively.

Collection
- **Selecting Seeds:** Choose seeds from strong, healthy plants with desirable traits—like great taste or the ability to thrive in your area.
- **Harvesting Seeds:** Allow fruits to fully ripen before extracting the seeds. For example, collect tomato seeds when the fruit is fully ripe or even slightly overripe.
- **Extracting Seeds:** Remove seeds from the fruits or pods, and clean them by removing pulp or debris. Rinse the seeds in water and let them dry thoroughly.

Drying
- **Air Drying:** Spread the clean seeds on a paper towel or screen in a warm place with good airflow, avoiding direct sunlight. Let them dry completely over a few days.
- **Checking for Dryness:** Test the seeds for dryness by attempting to bend one; if it snaps or breaks cleanly, they're ready!

Storage
- **Choosing Containers:** Use airtight containers like glass jars or resealable plastic bags to store the seeds. Ensure these are clean and dry, with no leftover moisture.
- **Labeling:** Label each container with the seed type, harvest date, and any other relevant information for future planting.
- **Cool, Dark Location:** Store the containers in a cool, dark place—like a pantry, cellar, or fridge. Avoid areas with significant temperature changes or excessive light, as these can affect seed viability.

COLLECTING AND PRESERVING FOOD

Harvesting Crops at Peak Ripeness

Picking your crops at just the right time is super important for getting the best taste and nutrition out of them. Here are some tips to ensure you're harvesting your produce when it's perfect:

- **Check Every Day:** Get into the groove of inspecting your garden daily. Look for signs like bright color, firmness, and size to know when things are ripe.
- **Early Bird Gets the Worm:** Try harvesting early in the day. Cooler temps help keep your crops fresh and prevent wilting.
- **Sharp Tools Rule:** Use sharp scissors or pruners to avoid damaging your plants and to make clean cuts.
- **Be Gentle:** Handle your veggies with care to prevent bruises or damage that could make them spoil faster.

Preserving Food Through Various Methods

Canning

Canning is an old-school way to keep fruits, veggies, and even meats in airtight jars. Here's how to do it step by step:

1. **Sterilize Jars:** Clean jars, lids, and bands with hot, soapy water, and boil them for ten minutes.
2. **Prep Your Produce:** Wash and prep your fruits or veggies. Remove any blemishes or bruises.
3. **Fill 'Em Up:** Pack your produce into sterilized jars, leaving the headspace as per recipe instructions.
4. **Process Jars:** Place filled jars in a water bath or pressure canner. Follow recommended processing times and pressures based on your altitude and food type.

Drying

Drying is a tried-and-true method that removes moisture from food to stop mold and germs from forming. Here's how to do it right:

1. **Pick the Right Stuff:** Choose high-moisture fruits, veggies, or herbs for drying, such as tomatoes, peppers, or herbs.
2. **Prep for Drying:** Wash and slice your produce uniformly. Arrange them on drying racks in a single layer.
3. **Dry Correctly:** Use a food dehydrator, oven, or air-drying method to gradually remove moisture from your produce.
4. **Store Dried Food Well:** Once dried, store the food in airtight containers in a cool, dark place to keep it fresh longer.

Freezing

Freezing is a quick and easy way to preserve garden goodies while keeping flavor and nutrients intact. Here's how to do it:

1. **Blanch First:** Briefly immerse veggies in boiling water, then quickly cool them in ice water to lock in color and texture.
2. **Proper Packaging:** Pack blanched or prepped produce into freezer-safe containers or bags, removing excess air to avoid freezer burn.
3. **Label & Date:** Write down what's inside and when you froze it on the containers for easy identification later.
4. **Freeze Fast:** Put filled containers directly into the freezer ASAP to maintain high quality.

Root Cellaring

Root cellaring is an old-school method for storing root veggies like potatoes, carrots, and onions in cool, humid spots. Here's how to do it:

1. **Pick a Spot:** Find a cool, dark area—maybe your basement? Make sure it has good ventilation too!
2. **Prep Storage Bins:** Clean storage bins or crates with drainage holes to avoid excess moisture buildup.
3. **Layer Those Veggies:** Layer different types of root veggies separately in bins so their ripening gases don't mix.
4. **Watch Temp & Humidity:** Keep an eye on temperature and humidity levels, and adjust as needed to maintain optimal storage conditions.

LONG-TERM SUSTAINABILITY: BUILDING RESILIENCE IN YOUR SURVIVAL GARDEN

By using permaculture principles and saving seeds, you can keep your garden thriving for the long term.

Permaculture Principles: Designing for Resilience

Permaculture is all about working with nature to create sustainable and regenerative systems. By following its principles, your garden becomes more resilient and productive.

- **Observation and Interaction:** Spend some time just observing your garden space. Notice how sunlight, wind, and water move through the area. Work with these natural processes instead of trying to fight them.
- **Design from Patterns to Details:** Start with the big picture, then move on to the details. Consider factors like plant groupings, microclimates, and where animals are likely to interact with your garden. Gradually refine your garden design as you go.
- **Use and Value Renewable Resources and Services:** Make the most of renewable resources like sunlight, rainwater, and organic matter. Set up rainwater harvesting systems, compost bins, and mulch to improve soil quality and retain moisture.

Seed Saving: Ensuring Future Harvests

Saving seeds from your own plants helps maintain diversity and resilience in your garden.

- **Selecting and Saving Seeds:** Choose open-pollinated or heirloom varieties of vegetables and herbs for seed saving. Allow them to fully mature before harvesting seeds from the best plants.
- **Cleaning and Storing Seeds:** Clean the seeds thoroughly to remove any debris or pulp. Ensure they are completely dry before storing them. Keep seeds in a cool, dry place, using airtight containers or envelopes to preserve their viability.
- **Testing Seed Viability:** Periodically perform germination tests on your stored seeds to check their viability. Place a few seeds on a damp paper towel and observe how many sprout. This will help ensure they are still good for future planting.

Incorporating Permaculture Guilds: Maximizing Diversity & Productivity

Permaculture guilds? They're basically plant groups that help each other grow. When you incorporate guild planting into your survival garden, you create super resilient ecosystems that need hardly any work.

- **Building Plant Guilds:** Okay, so first, pick a main crop. It could be a fruit tree or even a perennial veggie. Then, add companion plants around it. These will help with nutrients and keeping pests away. For variety, include nitrogen-fixing plants, dynamic accumulators, and ground covers. This mix boosts soil fertility and biodiversity.
- **Encouraging Beneficial Insect Habitats:** Want to see more helpful bugs like ladybugs, lacewings, or pollinators? Plant some native flowers, herbs, and flowering shrubs. These insects are awesome for controlling pests and pollinating plants. It really helps keep your garden healthy.

- **Mulching & Soil Building:** Use materials like straw, leaves, or wood chips as mulch. This helps retain moisture in the soil, suppress weeds, and improve soil quality overall. And don't forget about compost! Adding compost, compost teas, and green manures enriches the soil with loads of nutrients and beneficial microorganisms.

16. Hunting, Trapping, and Fishing

Hunting, trapping, and fishing are super important for preppers. These ancient practices have been around for thousands of years. They kept humans alive back then and still matter today, especially if things go bad.

A Look at Hunting, Trapping, and Fishing

Hunting, trapping, and fishing are the basics of surviving in the wild. Each has its own ups and downs, but together they give you a solid way to find food.

Hunting means using guns or bows to track down animals for food. You need skill, patience, and a good understanding of how animals act.

Trapping involves setting devices to catch animals without you being there. Tools like snares, deadfalls, and cage traps can catch both small and big animals.

Fishing can be done with a rod and reel or even simple tools like hand lines and nets. It's a steady way to get fish if you know where they hang out and how they behave.

Why Being Self-Sufficient Matters

Being able to get your own food means you'll be able to take care of yourself and your loved ones when things are tough.

Self-Reliance

Learning hunting, trapping, and fishing gives you the power to provide without needing outside help.

Security

These skills can help you deal with the worries of not having enough food during tough times.

Knowing How Animals Act Helps

Success in hunting, trapping, and fishing depends on understanding how animals behave. By studying their habits and where they live, you boost your chances of finding food in the wild.

Study of Animal Behavior and Habitats

Animals have their own special ways and places they like to hang out. Knowing these habits helps a lot with hunting and trapping game.

Tracking:

Following animals is like playing detective. You look for signs they leave behind, like footprints (tracks), poop (scat), and eaten plants (browse). Getting good at this means you can better guess where they're going next.

Habitat:

Animals live in certain kinds of places. By knowing where your target animals are likely to hang out, you can save time and effort by looking in the right spots.

Understanding Fish Habits and Habitats

Fish behavior changes with factors like water temperature, depth, and plants in the water. If you know these things, you can fish smarter.

Water Temperature:

Fish are cold-blooded, so the water temperature makes a big difference for them. They get sluggish in cold water but are much more active in warm water.

Habitat Features:

Fish love hiding around underwater structures like rocks, fallen trees, and beds of weeds. These spots offer good cover and places to catch prey. Fishing near these features boosts your chances of catching something.

SURVIVALIST HUNTING METHODS

Whether you're just starting out or you're a seasoned pro, getting the hang of the right hunting techniques can really boost your success. In this section, we'll chat about the key topics: picking and maintaining firearms, bowhunting basics and gear, tracking and stalking prey, and setting up blinds and stands.

Firearms Selection and Maintenance for Hunting

Firearms are super handy for hunters—think precision and power when you use them right. When you're picking a hunting firearm, consider factors like caliber, action type, and the game you're after. Many hunters choose rifles in calibers like .308 Winchester or .30-06 Springfield because they work well for various types of game.

After selecting your firearm, keeping it in good shape is a must. Clean and lubricate it regularly to prevent rust and breakdowns. Check the barrel, action, and magazine for any wear or damage, and replace any worn parts as needed. Store your firearm safely and keep it dry to maintain its condition over time.

Bowhunting Techniques and Equipment

Bowhunting is all about a simple and stealthy approach to hunting. When choosing a bow, consider factors like draw weight, draw length, and axle-to-axle length to ensure it fits just right. Compound bows are great for their power and accuracy, while traditional recurve bows offer simplicity and reliability.

Getting good at bowhunting takes practice and patience. Focus on improving your stance, grip, and aiming. Practice drawing and releasing the bow smoothly to minimize noise. Use high-quality arrows and broadheads matched to your bow to ensure effectiveness.

Tracking and Stalking Prey

Tracking prey requires keen observation and an understanding of animal behavior. Study tracks, scat (animal droppings), and other signs left by animals. Fresh signs, like disturbed plants or recent droppings, help you determine the animal's location.

When stalking prey, move slowly and quietly, using natural cover to conceal your movements. Pay attention to wind direction to avoid alerting animals with your scent. Stay low and minimize noise to remain hidden. Be patient; stalking often requires waiting for the perfect moment to strike.

Setting Up Blinds and Stands for Hunting

Blinds and stands provide cover and elevation, which can significantly boost your hunting success. Choose locations with high animal activity and clear shooting lanes when setting up blinds or stands. Conceal your blind with natural materials like leaves and branches so it blends in seamlessly.

For tree stands, ensure they are securely fastened to trees at a height that offers a clear line of sight. Use safety harnesses and lifelines to prevent falls. Also, practice shooting from your blind or stand to familiarize yourself with the angles and distances you'll encounter.

SURVIVAL TRAPPING TECHNIQUES

Trapping provides a cool and easy way to get food when you're out trying to survive. There are lots of ways to trap, allowing you to catch small animals like rabbits and squirrels or even bigger ones like raccoons and beavers. Let's explore some different techniques, tips for building your traps, and how to be kind to nature while you do it.

Overview of Different Trapping Methods

Snares:
Snares are pretty simple but effective. They basically stop an animal from moving around. A snare is a loop tied to something strong, like a tree or a stick. To set one up, find a path where animals often walk (or run) and place the loop at their head height or right on the trail. When the animal walks through it, the loop tightens around its neck or body, trapping it.

Deadfalls:
Deadfalls are old-school traps that use gravity to catch prey. Typically, a big rock or log is held up by some kind of trigger. When the trigger is tripped, the weight falls and crushes the animal underneath. Making a deadfall requires some skill and precise placement, but when done well, they work great.

Cage Traps:
Cage traps, also known as live traps, catch animals without harming them. They're made from wire mesh with a trigger that shuts the door once the critter goes inside. Cage traps can be used for a variety of animals, like raccoons, possums, and skunks. They're perfect for those who care about being humane while trapping and want the option to release any animals they didn't intend to catch.

Snares:

Construction: Okay, first find some durable wire or cord. You want it strong enough to catch the animal. Make a small loop at one end and secure the other end to something fixed, like a tree or stake.

Placement: Look for paths where animals often walk—these are great spots. Place the snare loop at head height or right on the trail. Hide it well, but make sure animals can still get into it. Try placing it near den entrances or feeding spots, too.

Deadfalls:

Construction: Pick a good base, like a flat rock or sturdy log. This will be your foundation. Use sticks or branches to make a trigger mechanism that's really sensitive. Then grab a heavy weight, like a big rock or log, to serve as the crushing part.

Placement: Set this trap in areas where your target hangs out, maybe by water sources or feeding spots. Put bait near the trigger so the animal walks right into it. Make sure everything is stable so the trap doesn't go off too soon.

Cage Traps:

Construction: Follow the instructions to put together the wire mesh enclosure securely. Attach the trigger mechanism, usually a door or gate, and ensure it works well. Test the trap to see if it opens and closes properly.

Placement: Pick places where your target animal often goes, like travel routes or near burrows. Use yummy bait, like peanut butter or fruits, to lure them in. Anchor the trap firmly so they can't escape once caught.

Ethical Considerations & Legal Regulations Regarding Trapping

Trapping is a handy skill, but it must be done responsibly and within the rules. Before setting any traps, familiarize yourself with local laws and guidelines. It's crucial to know what's allowed, which traps you can use, and when trapping is permitted.

Always consider the well-being of the animals. Focus on humane treatment and avoid causing unnecessary pain. To be ethical, try not to catch the wrong animals; if you do, release them unharmed. Check your traps frequently to ensure that no protected or endangered animals get caught. Quickly remove any unintended catches.

By adhering to ethical and legal standards, preppers can use trapping as a practical way to obtain food while respecting wildlife and maintaining ecological balance.

SURVIVAL FISHING TECHNIQUES

In this section, we'll dive into some handy fishing methods—like rod and reel tips, alternative fishing techniques, selecting the right bait, and staying safe while doing it.

Rod & Reel Fishing Techniques

For rod & reel fishing, being flexible is super important. It doesn't matter if you're by a calm lake, a busy river, or the sea. Knowing different techniques can up your game. Let's break it down:

Casting:
Grab your rod nice and easy. Place your dominant hand on the reel, and the other hand on the base. Push the button or open the bail (if using a spinning reel) to let the line go. Then, swing the rod back gently and quickly bring it forward to toss your bait or lure into the water.

Retrieving:
Got your bait in the water? Great! Now try different ways of bringing it back—steady reeling, little twitches, or jerks. These moves make your bait look like real prey and could tempt fish. Mix up your speed and rhythm to see what works best in different places.

Setting the Hook:
Felt a nibble? Don't yank the rod right away! First, reel in any loose line. Then, raise the rod tip smoothly but firmly to hook that fish in its mouth. Practice makes perfect—so you don't lose your catch.

Using Alternative Fishing Methods

Rod and reel fishing can be great, but sometimes knowing other ways to fish with just a bit of gear is super handy. Hand lines, trotlines, and nets? All super practical. Different settings need different approaches.

- **Hand Lines:** Simple yet clever. A hand line is just some fishing line wound on a spool or something similar. Attach a hook and bait, toss it in the water, then wait. They're small and light—perfect for hiking or survival.
- **Trotlines:** These are long lines with multiple hooks spaced out. Set them up at the bottom of a river or lake and leave them. Check back occasionally for fish. They're awesome because you can catch multiple fish at once without much effort.
- **Nets:** Fishing nets come in all shapes and sizes for shallow or deep waters. You can throw cast nets to trap fish or use dip nets to scoop them up. Nets make it easier to catch lots of fish quickly.

Understanding Bait Selection and Presentation

Picking the right bait is key to getting fish to bite. Different fish like different bait, so match your bait to the fish you're after.

- **Live Bait:** Worms, minnows, bugs—fish love these! Hook the bait just right so it looks natural and tempting.
- **Artificial Lures:** Lures come in many shapes, sizes, and colors to mimic real prey. Experiment with different ones to see what works best depending on where you're fishing and the conditions.

Safety Precautions and Regulations

Before you go fishing, always make safety your number one priority and follow the rules to keep yourself and nature safe.

- **Check Local Regulations:** Know your area's fishing rules—things like how many fish you can catch, size limits, and where you can fish.

- **Wear Protective Gear:** Dress appropriately for the trip! If you're on a boat or in deep water, wear a life jacket. Good shoes help prevent slips and falls.
- **Handle Fish Responsibly:** Be gentle with caught fish—use wet hands or a landing net to avoid harming them. Use catch-and-release methods if the fish are too small or if protected species need saving for their survival.

HANDLING AND STORING GAME AND FISH

Maximizing the benefits of your hunting and fishing trips is essential. Below, we'll cover the practical steps for **field dressing and butchering game**, **cleaning and filleting fish**, and **various methods for preserving meat and fish** during tough times.

FIELD DRESSING & BUTCHERING TECHNIQUES FOR GAME ANIMALS

Field dressing is crucial for preparing game animals for consumption. It involves removing the internal organs and cooling the carcass to prevent spoilage. Follow these steps:

1. Preparation

- **Position:** Lay the animal on its back.
- **Tools:** Gather a sharp knife, bone saw, and disposable gloves.

2. Start with the Incision

- **Incision:** Make a cut from the base of the sternum down to the pelvic bone.
- **Caution:** Be careful not to puncture the intestines.

3. Remove Internal Organs

- **Process:** Reach inside and carefully remove the heart, lungs, and other organs.
- **Additional Step:** Cut the windpipe and esophagus, then pull them out as well.

4. Butchering

- **Cutting:** Once the cavity is empty, start cutting the animal into manageable pieces.
- **Tools:** Use your bone saw for the legs and remove the backstraps.

5. Cooling

- **Importance:** It's crucial to cool the meat quickly.
- **Methods:** Hang the cuts in a shaded area or place them in a cooler with ice packs.

FILLETING & CLEANING FISH

Filleting and cleaning fish may seem tricky at first, but with practice, it becomes easier. Here's how to do it:

1. Preparation

- **Rinse:** Rinse your fish under cold water.
- **Dry:** Pat it dry with paper towels and place it on a clean cutting board.

2. Make the First Cut

- **Positioning:** Place a sharp fillet knife behind the gills.
- **Cut:** Slice downward towards the backbone.

3. Extract the Fillet
- **Technique:** Run your knife along the spine to free the fillet from the ribs.
- **Repeat:** Perform the same process on the other side of the fish.

4. Skinning
- **Scoring:** Carefully score the skin at the fillet's tail end.
- **Grip:** Ensure your knife is aimed toward the head while gripping the flesh firmly.

5. Cleaning
- **Final Check:** Inspect the fillets for any remaining bones or scales.
- **Removal:** Use tweezers or a scaler to remove any stragglers.

METHODS FOR PRESERVING MEAT & FISH

Preserving your food ensures it stays safe without refrigeration. Here are three reliable methods:

1. Smoking
- **Flavor:** Smoking adds flavor while extending shelf life.
- **Tools:** Use a barrel smoker or purchase one ready-made.
- **Temperature:** Smoke meat or fish at low temperatures (around 200°F) until fully cooked and flavorful.

2. Drying
- **Dehydration:** Drying removes moisture, preventing bacterial growth.
- **Preparation:** Thinly slice meat or fish.
- **Process:** Place in a dehydrator or oven set to a low temperature (around 140°F) until dry.

3. Canning
- **Process:** Canning involves sealing food in airtight jars processed in a pressure canner.
- **Guidelines:** Follow USDA-approved guidelines for safety.
- **Longevity:** Canned foods can last for years, offering convenient shelf-stable options.

SECURING EMERGENCY FOOD: ESSENTIAL FOR SURVIVAL

In times of crisis, having a steady food supply is essential. Adding hunting, trapping, and fishing to your emergency plans can provide a reliable source of food when store shelves are empty. Let's explore how to use wild game and fish as key food sources when necessary, along with tips to maximize these activities.

Adding Hunting, Trapping, and Fishing to Emergency Plans

When disaster strikes or society crumbles, typical food sources might be hard to find. That's when hunting, trapping, and fishing become invaluable for securing food. By including these in your emergency plans, you ensure a nutritious food supply for you and your loved ones.

How to Get the Most Food from Hunting, Trapping, and Fishing

- **Practice Conservation:** Only take what you need. Avoid waste to keep wild game and fish populations sustainable.
- **Use Different Tactics:** Try various hunting, trapping, and fishing methods to increase your chances of success in different environments and seasons.
- **Preserve Your Catch:** Properly process and preserve what you harvest to avoid spoilage and extend its shelf life. Techniques like smoking, drying, and canning can help maximize your food supplies.

- **Keep Getting Better:** Continuously improve your skills. Practice regularly. Educate yourself. Learn by doing. Stay updated on local regulations and conservation efforts to act responsibly toward nature.

SKILL DEVELOPMENT AND PRACTICE: MASTERING CRITICAL OUTDOOR SURVIVAL TECHNIQUES

Being good at hunting, trapping, and fishing can make all the difference. You could thrive instead of just surviving in a crisis. But how do you get really good? Regular training and practice.

Importance of Honing Skills Through Practice

Practice makes perfect. Doing things over and over sharpens instincts, polishes techniques, and builds confidence.

Practical Tips for Effective Training

- **Start Small:** Begin with the basics. Set up simple traps or cast a fishing line in the backyard.
- **Be Consistent:** Practice regularly. At least once a week is good—more as you improve.
- **Try Simulations:** Test your skills in real-life settings. Create mock hunting blinds or tie knots under pressure.
- **Get Feedback:** Ask experienced folks for advice. Hunters, trappers, or anglers can offer great tips.

Building Confidence and Proficiency in Outdoor Survival Skills

Know what? Mastery brings confidence. Here's how to boost yours:

- **Set Small Goals:** Break tasks into bite-sized chunks and celebrate each win.
- **Stay Persistent:** View setbacks as lessons and keep trying.
- **Be Aware:** Learn to read nature's signs—animal activities, tracks, and changing environments.
- **Enjoy the Journey:** Remember, mastery takes time. Love the learning process and enjoy growing your skills.

17. Sustainable Harvest

IMPORTANCE OF FORAGING FOR SURVIVAL

Foraging for wild edible plants is a skill that involves spotting and gathering nature's treasures, no fancy tools required. Here's why this skill is so crucial:

Self-Sufficiency:
Mastering the art of foraging makes you less dependent on modern comforts. You'll discover a newfound strength in being self-reliant. During emergencies, this knowledge is priceless.

Nutritional Diversity:
Wild plants are packed with vitamins, minerals, and nutrients you won't easily find on store shelves. Including these natural goodies in your meals means a richer and more varied diet.

Emergency Preparedness:
Food supply chains can be unreliable—think natural disasters, unexpected chaos, or economic instability. Being able to forage provides an essential fallback to keep you and your family nourished.

Cost-Effective:
Foraging is often free! It's a smart way to add nutrition to your meals without spending a dime. So, even during calm times, get outside and harvest some healthy bites from Mother Nature herself.

RECOGNIZING COMMON EDIBLE WILD PLANTS

Leaves
- **Observation:** Start by looking at the shape, color, texture, and how the leaves are arranged.
- **Field Guide:** Use a good field guide or a foraging book to spot common wild edible plants.
- **Key Features:** Notice special traits like toothed edges, lobed shapes, or unique veins.
- **Touch Test:** Gently feel the leaves. Check for prickles, spines, or other defensive features.
- **Smell:** Some edible leaves have a distinct smell, which helps in identifying them.
- **Caution:** Steer clear of leaves with milky sap, thorns, or a bitter taste, as they might be toxic.

Roots
- **Visual Examination:** Look closely at the size, shape, color, and texture of roots.
- **Pull Test:** Gently pull the plant from the dirt to examine the root structure.
- **Tuberous Roots:** Find tuberous roots that resemble potatoes; these can be dug up and eaten.
- **Taproots:** Spot plants with long taproots, typically found in dry or sandy soil.
- **Fibrous Roots:** Some edible plants have fibrous roots that can be gathered and cooked.
- **Underground Stems:** Plants like cattails have edible rhizomes or underground stems that you can dig up and eat.

Berries
- **Observe the Color, Shape & Arrangement:**
  First, look closely at the color, shape, and how the berries are arranged on the plant.
- **Use a Reliable Field Guide:**
  Grab a reliable field guide or a trusty foraging app to identify common wild berries. These tools are valuable.
- **Look for Cluster Formation:**
  Notice if berries grow in clusters or specific patterns. This can be a helpful clue.
- **Conduct a Taste Test:**
  Taste a small part of the berry to check for bitterness or unpleasant flavors. Don't eat a large quantity until you're sure they're safe.
- **Learn Fruit Anatomy:**
  Understand the differences between berries, drupes, and other types of fruits. This knowledge helps avoid toxic varieties.
- **Be Cautious:**
  Steer clear of berries that taste bitter, have thorns, or smell bad. They might be poisonous.

Mushrooms
- **Do a Visual Inspection:**
  Carefully examine the size, shape, color, and texture of the mushroom.
- **Consult a Field Guide:**
  Refer to a reliable field guide or foraging app to confidently identify wild mushrooms.
- **Note Growth Habit:**
  Pay attention to where and on what the mushroom is growing. This can offer valuable clues.
- **Analyze Cap and Stem:**
  Examine the mushroom's cap and stem—pay attention to gills, pores, or spines. Every detail counts.
- **Take a Spore Print:**
  Place the mushroom cap on paper to observe its spore color. It's an essential identification step.
- **Perform an Edibility Test:**
  Cook a small piece of the mushroom and then wait 24 hours to check for any adverse reactions.

- **Exercise Caution:**
  Avoid mushrooms that have a bad smell, are slimy, or show signs of decay. They could be poisonous.

NUTRITIONAL BENEFITS AND HEALING PROPERTIES

Understanding the nutritional value and properties of wild edible plants is crucial. These plants not only provide sustenance but also offer health benefits that are invaluable in survival situations.

Nutritional Value:

- **Vitamins and Minerals:**
  Wild edible plants often brim with essential vitamins and minerals crucial for optimal health. For example, dandelion greens are rich in vitamin A, vitamin C, and calcium, while watercress boasts high levels of vitamin K and vitamin C.
  To ensure you get diverse nutrients, include various wild plants in your diet.

- **Fiber:**
  Many wild plants are loaded with dietary fiber, which aids digestion and helps regulate blood sugar levels. Chickweed and purslane, for instance, are high in fiber.
  Add fiber-rich plants like dandelion greens, chicory, and burdock to your meals for better digestive health.

- **Protein:**
  Though wild plants may not match animal sources in protein content, some varieties offer significant amounts, making them valuable for a prepper's diet. Amaranth seeds and chickweed are examples of protein-rich wild plants.
  Seek out protein-rich plants like amaranth, chickweed, and watercress to supplement your protein intake in the wild.

Medicinal Properties:

- **Anti-inflammatory:**
  Several wild plants have anti-inflammatory properties, helping to alleviate pain and swelling from injuries or illnesses. Plantain leaves and fireweed are known for their anti-inflammatory effects.
  Consider adding plants like plantain, chickweed, and fireweed to your diet to benefit from their anti-inflammatory properties.

- **Antioxidants:**
  Many wild plants are rich in antioxidants, which protect cells from damage caused by free radicals. Elderberry and rose hips have potent antioxidant properties.
  Consume antioxidant-rich plants like wild mint, wood sorrel, and purslane to support overall health and well-being.

- **Immune Support:**
  Some wild plants boost the immune system, making them great allies against infections and illnesses. Garlic mustard and elderberry are known for their immune-supportive qualities.
  Include immune-supportive plants like elderberry, rose hips, and garlic mustard in your diet to strengthen your body's defenses.

POSSIBLE DANGERS OF WILD PLANTS

Wild edible plants can be beneficial, but you need to be aware of the dangers they can bring. Eating the wrong plant or not handling it properly could make you sick—or worse. This makes proper identification and care incredibly important when you're foraging.

Toxicity:
Many wild plants contain toxins that can cause serious health issues if consumed in large quantities or if not prepared correctly. For example, certain mushrooms, like the death cap, are notoriously deadly. Always do your homework—research and correctly identify any wild plant before consuming it to avoid toxicity issues.

Allergies:
Some people might be allergic to certain wild plants. Reactions can range from mild itching to severe health problems. Plants like ragweed and poison ivy are common culprits. If you know you're sensitive, be cautious when trying new wild plants. Start small and observe how your body reacts.

Contamination:
Wild plants can come into contact with pollutants, pesticides, or other harmful substances. For instance, plants growing near factories or farms might have high toxin levels. Choose your foraging spots wisely—avoid areas near industrial sites, farm fields, and busy roads where contamination is likely.

Misidentification:
Confusing a toxic plant with an edible one is a common mistake in foraging. For example, poison hemlock resembles wild carrots but is extremely toxic. Use reliable field guides, check online resources, or consult experienced foragers to ensure you're correctly identifying plants before consuming them.

RESPONSIBLE FORAGING GUIDELINES

Follow these essential tips for a safe foraging experience and minimize risks:

Proper Identification: Always ensure you know which wild plants you are picking. Use more than one trusted source, such as field guides and foraging apps, to confirm each plant's identity. Pay close attention to key features like leaf shape, flower type, and the plant's natural habitat.

Harvesting Techniques: When gathering plants, only take from areas where they are abundant, ensuring you don't deplete the environment. Use clean, sharp tools for cutting and leave enough leaves behind for the plant to regrow, allowing animals to continue relying on those plants as well.

Preparation Methods: Some plants require specific preparation methods to remove bitterness or toxins, making them safer and more palatable. Techniques like boiling, blanching, or fermenting can be crucial. Always follow the recommended preparation methods for each type of plant.

Environmental Awareness: Be a mindful forager by respecting nature. Avoid disturbing animals and make sure to leave no trace behind. Adhere to local regulations and support efforts to protect wild plants and their natural habitats.

SEASONAL PRESENCE AND HABITAT CHOICES

Different wild edible plants pop up in different seasons and locations. If you can identify these factors, you'll have a much better shot at finding a variety of food.

Spring:
In spring, wild edible plants start to sprout as the cold winter fades away. Look out for nettles, dandelions, and plantains. These are often the first greens you'll see. They thrive in moist areas with plenty of sunlight, such as fields, forests, and riverbanks.

Summer:
As the weather heats up, more wild edibles appear. Berries like blackberries, raspberries, and blueberries begin to ripen. Also, keep an eye out for wild mint and wood sorrel. These plants can be found in various locations, including woodlands, meadows, and roadsides.

Fall:
Fall is prime time for nuts and seeds! Look for acorns, chestnuts, and hazelnuts. Mushrooms also become abundant during this season, as many types thrive in the cool, damp autumn air. Search in wooded areas, orchards, and parks.

Winter:
Pickings are slimmer in winter, but there are still some options! You can make tea from evergreen needles, which are packed with vitamin C. Some root vegetables, like burdock, can still be dug up even under the snow. Focus your search on sunny slopes and wind-protected valleys for these hardy plants that endure through winter.

AWARENESS AND SUSTAINABILITY FOR THE ENVIRONMENT

Foraging for wild edible plants ties directly into environmental awareness and sustainability. Understanding how our actions impact the ecosystem is crucial, and we must forage in ways that preserve both the environment and the plants' health.

Leave No Trace:

Take only what you need, leaving the rest for wildlife and future foragers. Avoid harming nearby vegetation or habitats.

Avoid Overharvesting:

Be mindful of the size and growth rate of the plants you're picking. Don't overharvest; allow enough to remain so they can continue growing and sustaining their populations.

Respect Wildlife:

Remember, wild edible plants are not just for us; they're for wildlife too. Avoid disturbing wildlife habitats or foraging in areas where rare or endangered species live.

Learn Responsible Harvesting Techniques:

Use proper harvesting methods to minimize plant damage. Employ sharp tools like knives or scissors for clean cuts, and avoid uprooting entire plants.

Cultivate a Relationship with Nature:

Foraging isn't just about finding food; it's about connecting with nature and appreciating its richness. Spend time learning about plants and their ecosystems.

18. Firearms and Ammunition

Firearms are vital for self-defense, hunting, and survival during tough times. They're also useful for bartering in post-collapse scenarios. And ammo? Super valuable.

Let's dive in deeper now. We've mentioned firearms before in other chapters, but this one? It's all about how to use and maintain your firearms just right.

We'll cover everything needed to make the most of this tool. It's not just about using them effectively but also about keeping you and your loved ones safe from misuse. So get ready to learn!

Principles of Firearm Safety & Proficiency

Safety First: Learn how to use a firearm safely. Never, ever point it at something you don't plan to shoot. Always treat every firearm as if it's loaded.

Know Your Firearm: Understand your firearm's unique traits and how it works. Learn its operation, handle it securely, and know how to load and unload it properly.

Practice Proper Storage: Keep firearms out of unauthorized hands—store them in a locked cabinet or safe. Store ammunition and firearms separately to avoid accidents.

Regular Maintenance: Clean your firearms regularly and keep them in top shape. Frequently check for any signs of wear or damage.

Seek Training: Proper training is key. Enroll in a top-notch firearms course to learn essential skills like marksmanship, handling, and emergency procedures.

Stay Informed: Stay up-to-date with local firearm laws and regulations. Knowing legal requirements is crucial for responsible ownership.

COMPREHENDING AMMUNITION

Overview of Different Calibers & Types

Caliber: This refers to the diameter of a bullet or the inside diameter of a firearm's barrel. It's vital to match the caliber of your ammunition with your gun's caliber to ensure safe and effective shooting.

Types of Ammunition:

- **Handgun Ammunition:** Used in pistols and revolvers. Comes in various calibers like 9mm, .45 ACP, and .38 Special. Designed for short to medium-range shooting. Commonly used for self-defense, target practice, and recreational shooting.
- **Rifle Ammunition:** Made for long-range shooting. Common calibers include 5.56mm, .308 Winchester, and .30-06 Springfield. Used in rifles and carbines for hunting, precision shooting, and military roles.
- **Shotgun Ammunition:** Shotgun ammo includes shotshells loaded with pellets (birdshot, buckshot) or a single slug. Shotguns are versatile tools suitable for sport, home defense, and hunting.
- **Rimfire vs. Centerfire:** Ammunition can be classified based on primer location. Centerfire ammunition has the primer in the center of the base, while rimfire has it in the rim of the case. Overall, centerfire is more reliable and powerful.

Factors to Consider When Choosing Ammunition

- **Firearm Compatibility:** Always ensure your ammunition matches your firearm's caliber. Using the wrong type can damage your firearm or cause malfunctions.
- **Purpose:** Consider why you're using the ammunition—whether for target practice, hunting, or self-defense. Choose based on your specific needs.
- **Bullet Weight and Design:** Bullets come in different weights and shapes, affecting their performance and trajectory. Lighter bullets travel faster but may not penetrate as well, while heavier bullets penetrate better but might lose speed. Choose what fits your goal.
- **Terminal Ballistics:** This refers to how a bullet behaves when it hits its target. Consider expansion, penetration, and energy transfer when choosing ammunition for self-defense or hunting.
- **Cost and Availability:** Consider the cost and availability of ammunition, especially if you plan to stock up or shoot frequently. Premium ammunition performs well but costs more and may be harder to find. Balance quality with affordability based on your needs and budget.

FIVE MUST-HAVE FIREARMS FOR PREPPERS

Firearms are crucial for protection, sustenance, and versatility in tough times. In this section, we'll dive into the top five essential firearms every prepper needs. From compact pistols for personal defense to powerful rifles for hunting and long-range engagements, each gun has a specific role in a prepper's toolkit. Understanding the functions and capabilities of these firearms is vital for effective readiness in any situation.

1. Semi-Automatic Pistol in 9mm

- **Purpose:**
  The 9mm semi-automatic handgun is a must-have due to its adaptability and reliability in self-defense situations. It's perfect for daily use because of its small size, making it easy to carry and conceal. This pistol offers you and your loved ones a line of defense in emergencies.

- **Function:**
  Insert a loaded magazine into the pistol, then rack the slide to chamber a round. Aim at your target and squeeze the trigger to fire. For swift follow-up shots, the semi-automatic action ejects the spent case and loads the next round immediately.

2. Carbine Rifle in 5.56 or .223

- **Purpose:**
  The carbine rifle chambered in 5.56 or .223 is versatile, providing effective performance in both short-range and medium-range engagements. Its lightweight design and manageable recoil make it suitable for hunting, defense, and tactical scenarios.
- **Function:**
  Load a magazine into the rifle and charge it by pulling the charging handle. Aim at your target and fire by squeezing the trigger. The semi-automatic action ensures quick and precise shooting, an essential trait for preppers in multiple situations.

3. 12-Gauge Pump Shotgun

- **Purpose:**
  The 12-gauge pump shotgun is a cornerstone firearm, valued for its significant stopping power and adaptability. It excels in close-quarter confrontations and is highly effective for hunting game or defending against threats.
- **Function:**
  Load shells into the magazine tube and pump the action to chamber a round. Aim at your target and pull the trigger to fire. The pump action ejects the spent shell and chambers another one quickly, providing reliable firepower when needed.

4. .22 Caliber Rifle

- **Purpose:**
  The .22 caliber rifle is practical and economical, offering versatility in hunting small game, target shooting, and training. Its low recoil and quiet operation make it an ideal gun for both beginners and seasoned shooters.
- **Function:**
  Load a single round or magazine into the chamber. Aim at your target, then press the trigger to fire. The rifle's bolt action or semi-automatic action cycles the next round for continuous shooting, giving preppers a trusty tool for gathering food and improving marksmanship skills.

REAL-WORLD USES OF FIREARMS IN PREPAREDNESS

Home Defense Strategies

When thinking about keeping your home safe, it's important to have a strong plan. Firearms can be a big help:

- **Firearm Selection:** Pick a trusty 12 Gauge Pump Shotgun for home defense. It's versatile and packs a punch, which is great for close encounters.
- **Training and Familiarization:** Make sure you practice with your shotgun regularly. Get comfortable with loading, aiming, and firing, especially when under stress.
- **Safe Storage:** Your shotgun should be easily accessible but also secure. Use a good gun safe or a quick-access lockbox.
- **Home Security Measures:** Don't rely solely on your firearm. Incorporate other security measures like strong doors, alarm systems, and motion-activated lights to keep intruders at bay.

Hunting and Food Procurement Techniques

In survival scenarios, securing food is crucial. Firearms can greatly assist in hunting:

- **Firearm Selection:** Use a .22 Caliber Rifle for small game hunting. It has low recoil and is affordable, making it perfect for frequent hunting trips.
- **Ammunition Considerations:** Stock up on plenty of .22 caliber ammo since it's light and easy to carry.
- **Tracking and Scouting:** Learn basic tracking and scouting skills to improve your ability to locate game animals.
- **Shot Placement:** Practice shooting targets to improve your accuracy. Aim for vital organs to ensure a quick and humane kill.
- **Field Dressing:** Be proficient in field dressing (cleaning) and processing game animals to avoid wasting any meat.

Bartering and Trade with Ammunition

After a major disaster, ammunition might become highly valuable for trading. Here's how to use your ammo wisely:

- **Ammunition Types:** Gather a variety of ammo types, such as 9mm, 5.56, and 12 gauge shells—these are popular and in high demand.
- **Establishing Value:** Determine the value of your ammo based on scarcity and demand. Keep an eye on trends to ensure you get the best deals.
- **Trading Strategy:** Be strategic when trading ammo for necessities. Be prepared to negotiate and aim for mutually beneficial deals.
- **Maintaining Discretion:** Be cautious about discussing your ammo stockpile with others. Keeping your stash a secret reduces the risk of theft or exploitation.

ENHANCED FIREARM TACTICS FOR PREPPERS

Customization and Modification of Firearms

Purpose:

Customizing and tweaking your firearms helps preppers make their weapons just right for specific tasks. It boosts their effectiveness and versatility in different situations.

Upgrade Your Sights:

Swap out factory sights with high-visibility options like fiber optic or tritium night sights.

This improves accuracy and makes targets easier to see, even in low-light conditions.

Install a Quality Trigger:

A smoother, lighter trigger pull can significantly enhance shooting control and accuracy.

Consider using drop-in trigger kits; they are easy to install and customize.

Add Tactical Accessories:

Attachments like weapon lights, lasers, and foregrips can improve handling.

Use rail systems to easily add or remove accessories as needed.

Tactical Considerations for SHTF Scenarios

Purpose:

Tactical planning is crucial for preppers to use firearms effectively in survival situations. This maximizes defensive capabilities and ensures a strategic advantage.

Practice Situational Awareness:

Stay alert and constantly observe your surroundings.

Spot potential threats ahead of time and have escape routes ready.

Implement Cover and Concealment:

Use cover to protect yourself from incoming fire.

Conceal your position to avoid detection by hostile forces.

Develop Communication Protocols:

Set up clear communication methods to coordinate movements and actions.

Use hand signals or predetermined signs for silent communication in high-risk scenarios.

Community Building and Mutual Support Networks

Purpose:

Creating a community with like-minded individuals facilitates collaboration, resource sharing, and collective defense. This strengthens overall preparedness and resilience.

Join Prepper Groups or Forums:

Connect with other preppers online or locally to share knowledge and resources.

Participate in group training exercises and drills to enhance skills and build camaraderie.

Establish Mutual Aid Agreements:

Form alliances with prepper neighbors for mutual support during crises.

Share supplies, skills, and manpower to boost collective survival efforts.

Organize Community Defense Initiatives:

Work together with fellow preppers to develop community defense plans and strategies.

Conduct regular patrols and perimeter checks to prevent threats and maintain security.

PROJECT 20: STORE AMMO FOR THE LONG HAUL

Storing your ammo is going to take more than just put it in a drawer, something that every responsible gun owner knows. However, there's a bit more to it if you are going through a natural disaster or other situation that seriously disrupts everyday life.

This section provides tips for ensuring your ammo remains useful.

Most ammo has a shelf life of 10 years. However, when stored securely, they can last longer.

Materials and Preparation

Before you start, you will need the following materials:

- Ammo boxes

Steps

1. When you buy ammo, write the date when you got it and keep a tally of how much ammo is in each box. If you leave the ammo in your house, the original box should be adequate.
2. If you take ammo outside, store the ammo in a military grade ammo box. You can leave it in the original box, making sure you keep updating the information on the outside as you use ammo.
3. Once you take ammo out of your home, do not return it to the stockpile. You have several options for storing it.
    a. Vacuum seal it to prevent rust and other issues.
    b. Use silica packets to keep them dry.
    c. Regularly check the boxes that have been removed from the home.

Ammo Box

PROJECT 21: RECYCLE AMMO AT HOME: STEP-BY-STEP GUIDE

Like most things, ammo is recyclable if you no longer need it or it is unusable.

Materials and Preparation

Before you start, you will need the following materials:

- Ammo Press
- Reloading Die
- Safety glasses
- Gloves
- Clean area to work
- No source of heat (such as fire or space heater)
- Reloading scale
- Ammo reload instructions
- Ammo and die components for each type of ammo you have

Steps

1. Make sure you have the following details for the specific ammo you plan to put in an ammo press.
    a. Bullet grain and type
    b. Calibur
    c. Charge amount
    d. FPS
    e. Powder type
2. Follow the instructions in the ammo press to ensure you follow the process set by the manufacturer. Different ammo have different needs, and this process can be fatal when not done exactly as the process in the tool's manual specifies.
3. You can also install dies and prepare charges, but all of these should be learned with an expert to ensure you don't accidently set off your ammo. Get classes and purchase the equipment specified by your instructor so that you have practice in a secure setting before you try it yourself.

Ammo

PROJECT 22: HOW TO MAKE BLACK POWDER SAFELY

Black powder has been in use for centuries, so it is something that has been around long enough for people to figure out how to make it on their own.

Be very careful and make sure you are in a clean, secure environment to do this project.

Materials and Preparation

Before you start, you will need the following materials:

- Potassium nitrate
- Charcoal ingredients
  - wood chips such as birch, beech ash, pine, and spruce
- Large pot
- Sulfur
- Seive (not used for food)
- Mortar and pestle
- Isopropyl alcohol
- Refrigerator
- Water
- Cheesecloth

Make sure the ingredient you get are of the highest quality that you can. This will make better powder.

Steps

1. Put the wood chips in the large pot or a large barrel with a lid. There needs to be a hole so steam rises, allowing the wood to char.
2. Heat up the pot or barrel. Remove it when the wood has charred, making charcoal.
3. Determine how much of each ingredient you want to use, keeping the following ratio in mind as you calculate the amount.
   a. 75% potassium nitrate
   b. 15% charcoal
   c. 10% sulfur
4. Put 1.5 cups of isopropyl alcohol for every 100 g or sulfur/charcoal you plan to mix.
5. Use the mortar and pestle to grind potassium nitrate, then put the contents to the side.
6. Use the mortar and pestle to grind the charcoal until it is a fine powder, then put the contents to the side.
7. Use the mortar and pestle to grind the sulfur into a fine powder, then put the contents of the side.
8. Measure out the proper ratio of each ingredient, basing the ration on the weight of each component.
9. Add 0.25 cup of water for every 0.5 cup of potassium nitrate you plan to add. Combine them in an old pan you *will not use* for food.

Adding Water and Potassium Nitrate

10. Boil the mixture, continually stirring it and adding small amounts of water. Stop only when all of the potassium nitrate dissolves.
11. Add the charcoal and sulfur and stir the three components together.
12. Get the chilled alcohol and your new mixture and head outside.
13. Get to safe place, then carefully pour the new mixture into the chilled isopropyl alcohol.
14. Stir the new solution.
15. Put the solution into the fridge to chill it.
16. Strain out the liquid by placing the chilled mixture on the cheesecloth. Have a bucket or other container to catch the liquid, then dispose of the liquid somewhere safe.

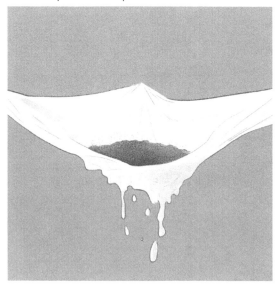

Add image of the bottle full of the three layers.

17. Put the mixture out in the sun to dry more thoroughly. Ensure it is secure and not easily blown away by a breeze or a knock on a table where it is resting.
18. When still slightly damp, put the mixture through a sieve, then place it back in the sun.
19. Run it through a sieve again several times to ensure that it is fully broken up.
20. Put the powder in a secure plastic container, then place it in a cool dry place.

Make sure you keep the final product well out of reach of children and pets.

19. Homesteading Projects

OFF-GRID LIVING: THE ULTIMATE STATEMENT OF INDEPENDENCE

Key Components of Off-Grid Living

1. **Energy Independence:**
   Generate your own electricity from renewable sources like solar, wind, or hydroelectric power.
2. **Water Self-Sufficiency:**
   Control your water supply by drilling wells, collecting rainwater, or using other eco-friendly methods.
3. **Food Production:**
   Grow your own food through gardening, raising animals, or by hunting and foraging.
4. **Waste Management:**
   Recycle and compost to reduce waste and minimize your environmental impact.
5. **Security and Defense:**
   Protect your homestead with smart planning and preparedness. Safety first!
6. **Communication & Connectivity:**
   Establish alternative communication channels for emergencies and stay connected with the outside world.

7. **Financial Sustainability:**
   Manage your finances wisely and create income streams to support your off-grid lifestyle.

Benefits of Off-Grid Living for Preppers

- **Self-Reliance and Independence**
  Off-grid living empowers you to be self-reliant. You can meet your basic needs without relying on external infrastructure or resources.

- **Resilience in Times of Crisis**
  Having your own sources of power, water, and food allows you to withstand grid disruptions and maintain a sense of normalcy and security.

- **Environmental Sustainability**
  Living off the grid reduces your carbon footprint and dependence on fossil fuels. This supports environmental health and contributes to a better future by using renewable energy and sustainable practices.

- **Cost Savings**
  While off-grid infrastructure requires a significant initial investment, the long-term savings can be substantial. By growing your own food and generating your own power, you can drastically reduce—or even eliminate—monthly utility bills, saving a significant amount of money.

- **Freedom and Peace of Mind**
  Off-grid living offers unmatched freedom and peace of mind, giving you control over your life and a sense of security.

KEY FACTORS TO KEEP IN MIND FOR OFF-GRID LIVING

Picking the Right Spot

The first thing you need to do to live off-grid is to find the perfect location for your new home. Here's how you can go about it:

- **Check Sunlight:** Ensure your spot gets plenty of sun all year round. Use a solar calculator or app to assess sunlight exposure. Look for areas with minimal shade from trees or buildings to maximize the efficiency of your solar panels.

- **Find Water Sources:** You'll need a reliable water source, such as wells, springs, or rivers. Investigate the groundwater in your area and consult local experts about water depth and quality.

- **Think About Weather:** Research the local climate, including temperature highs and lows, rainfall patterns, and seasonal changes. Choose a location where you'll be comfortable and able to grow food effectively.

- **Ease of Access:** Consider proximity to essential services like food supplies, hospitals, and emergency assistance. Being close to towns can be beneficial, but balance this with your desired level of isolation.

Legal Stuff

Understanding local laws and regulations for off-grid living is crucial to avoid potential issues. Here's what you need to consider:

- **Zoning Laws:** Check if off-grid living is permitted in your area by reviewing zoning regulations and building codes. Determine whether alternative housing types and sustainable practices are allowed.

- **Get Permits:** Obtain all necessary permits for construction, water use, and land management. Contact local government offices to ensure you meet all requirements.

- **Environmental Rules:** Familiarize yourself with regulations concerning land use, waste management, and natural resource conservation. Adopt eco-friendly practices to comply with these rules and protect the environment.

- **Legal Precedents:** Research previous legal cases related to off-grid living in your area. Consult legal experts or experienced off-gridders to help you navigate any legal challenges.

Resource Planning

Proper resource planning is key to sustainable off-grid living. Here's a simple guide:

- **Energy Needs:** Calculate your daily power requirements. Based on this, choose renewable energy systems like solar panels, wind turbines, or other options that best suit your needs.

- **Water Needs:** Estimate your water usage for drinking, cooking, irrigation, and animal care. Consider rainwater harvesting or drilling a well to ensure a consistent water supply.

- **Food Production:** Plan your food production by evaluating your land for gardening and livestock grazing. Consider soil quality, climate conditions, and seasonal crop growth.
- **Waste Management:** Develop an effective waste management plan that emphasizes recycling and composting. Implement sustainable practices to minimize pollution and conserve resources.
- **Emergency Plans:** Prepare for emergencies by stocking supplies and creating evacuation plans. Stay informed about local risks like wildfires or floods and establish reliable communication methods for these events.

CREATING YOUR SELF-SUFFICIENT HOMESTEAD

Sustainable Building Materials

When you're gearing up to build your off-grid homestead, picking the right sustainable materials is key. You want something that's good for the environment and lasts long. Here's the scoop on some great options:

- **Timber:** Go for locally sourced timber from forests that are managed sustainably. Timber is versatile, renewable, and biodegradable. It's great for framing, flooring, and siding.
- **Recycled Materials:** Try repurposing salvaged items like bricks, metal roofing, and lumber from old buildings. This reduces waste and adds a unique charm to your place.
- **Natural Insulation:** Use insulation materials like sheep's wool or recycled denim. They're thermally efficient and eco-friendly, keeping your home comfy all year while cutting down on energy use.
- **Earth and Cob:** Build with earth and cob (a mix of clay, sand, and straw) for walls and floors. These materials are plentiful and affordable. Plus, they have excellent thermal mass properties that help regulate indoor temperatures naturally.

Energy-Efficient Design Principles

Using energy-efficient design principles in your off-grid homestead can do wonders for reducing your environmental footprint. Plus, it'll save you some money! Here's what to keep in mind:

- **Passive Solar Design:** Position your home to soak up the sun's rays during winter while avoiding too much heat in the summer. South-facing windows can let sunlight warm up your home during the day, reducing the need for artificial heating.
- **High Thermal Mass:** Opt for materials with high thermal mass like concrete, stone, or adobe. They absorb heat during the day and release it slowly at night, keeping indoor temperatures steady.
- **Energy-Efficient Appliances:** Get ENERGY STAR-rated appliances like fridges, washing machines, and water heaters to cut down on energy use. They're designed to work more efficiently and will reduce your need for off-grid power sources.
- **Proper Insulation:** Ensure your home is well-insulated to prevent heat loss in winter and heat gain in summer. Use insulating materials like spray foam, cellulose, or rigid foam boards in walls, ceilings, and floors to create a thermal barrier.
- **Off-Grid Power Solutions:** Incorporate renewable energy sources such as solar panels and wind turbines into your design. These tap into nature's power to generate electricity, providing a reliable and sustainable energy source for your homestead.

ALTERNATIVE POWER SOURCES FOR OFF-GRID LIVING

Solar Power Systems

As we discussed earlier, solar power systems are a dependable and sustainable energy source for living off-grid. Here's what you need to know about these power systems:

- **Solar Panels:** Place solar panels on a surface facing south where they get the most sunlight. These panels use photovoltaic cells to capture sunlight and convert it into electricity.
- **Charge Controller:** A charge controller, placed between the battery bank and solar panels, regulates voltage and prevents overcharging. This device ensures your batteries charge efficiently and last longer.
- **Battery Bank:** Store extra energy generated by the solar panels in a battery bank. This allows you to use it at night or during cloudy days. Opt for deep-cycle batteries designed specifically for renewable energy.
- **Inverter:** An inverter is necessary to convert the DC electricity from the battery bank into AC electricity, which powers household appliances and gadgets.

- **Wiring and Circuit Protection:** Proper wiring and circuit protection devices are crucial. They connect the solar panels, charge controller, battery bank, and inverter safely and efficiently.

Wind Power Systems

In areas where the wind blows consistently, wind power systems can be a great supplement or even a replacement for solar power. Here's how to set it up for your off-grid home:

Wind Turbine:

Install a wind turbine in a wide-open space. The blades capture the wind's energy and convert it into rotational motion.

Charge Controller & Battery Bank:

Just like with solar power, you'll need a charge controller and a battery bank to store energy from the turbine. This ensures you have electricity even when there's no wind.

Inverter:

An inverter is necessary to convert the DC power from the battery into AC power for home use. Ensure that your turbine matches the power and voltage specifications of your inverter.

Tower and Mounting System:

Anchor your tower firmly into the ground. Strong winds won't be a problem if it's stable. The height and location are crucial—choose wisely to catch as much wind as possible.

Maintenance:

Regular checks and maintenance are essential to keep your system running smoothly. Monitor for wear on the blades, the tower, and electrical components.

Hydroelectric Power Systems

If you have a stream or river nearby, hydroelectric power systems can provide a steady, reliable source of energy. Here's how to set one up:

Water Turbine:

Place a water turbine in flowing water to harness energy. Choose a turbine that suits your water source's flow rate and head (the vertical drop).

Generator & Control System:

Connect the water turbine to a generator and control system. This converts the rotational motion into electricity. Make sure the generator size matches your energy needs.

Transmission & Distribution:

Use wires and panels to transmit electricity from the hydro system to your home. Consider the distance and potential voltage drop when setting this up.

Environmental Considerations:

Before installation, assess how it will impact the local ecosystem and obtain any necessary permits. Ensure that the design and operation are environmentally friendly.

WATER CONSERVATION AND COLLECTION

Rainwater Collection Systems

We've covered a lot about water management before, but let me refresh your memory. Harnessing nature's power to supply your off-grid homestead is not just sustainable—it's essential for true self-sufficiency.

With rainwater collection systems, you can capture and store rainwater for various uses. Here's a step-by-step guide to get you started:

1. Gutter Installation:

First, install gutters along the roofline of your home. This channels the rainwater to designated collection points. Make sure the gutters slope towards downspouts for proper water flow.

2. Downspout Diversion:

Next, direct the water flow from downspouts into rain barrels or cisterns for storage. Place these containers on stable ground and secure them with a sturdy base to prevent tipping over.

3. Screen Filters:

Install screen filters at the entry points of your gutters and downspouts. This step keeps debris like leaves, twigs, and insects out of your system. Clean and maintain these filters regularly to ensure they perform well.

4. Overflow Prevention:

Include overflow outlets in your rainwater collection system. This prevents water from backing up and causing damage to your home's foundation. Direct excess water away from the foundation using perforated pipes or swales.

5. Water Storage Tanks:

Choose durable, food-grade storage tanks, such as polyethylene or fiberglass, to store collected rainwater. Place the tanks in a shaded area to minimize algae growth and evaporation.

Well Digging and Maintenance

If you're living off the grid and can't rely on city water, digging a well is a great way to secure your water supply. Here's how to get it done:

Site Selection: First things first, consult a professional to choose the best spot for your well. Consider factors like groundwater depth, soil conditions, and proximity to potential pollution sources.

Drilling Process: Hire a licensed well driller. They have the right tools to dig the well shaft properly. Keep an eye on the process to ensure they dig deep enough and that you get a good water flow.

Well Casing Installation: Next, install a well casing. Use a durable material like steel or PVC to prevent the well shaft from collapsing or becoming contaminated. Ensure there's a good seal around it to keep out any surface water contamination.

Water Quality Testing: Regularly test your well water to ensure it's safe to drink. Look for contaminants like bacteria, nitrates, and heavy metals. If any issues arise, address them immediately.

Well Maintenance: Keep your well in top condition with regular cleaning, disinfection, and check-ups. Monitor the water level and call in professionals if your pump needs repair.

Water Filtration and Purification Methods

Even though rainwater and well water come from nature, you still need to make sure they're clean for drinking and other household uses. Here's how:

Mechanical Filtration: Use filters to remove dirt, sediment, and organic material from your water. These could be sediment filters, carbon filters, or ceramic ones. Replace them regularly to maintain efficiency.

UV Sterilization: Install UV light systems to eliminate harmful bacteria, viruses, and protozoa in your water. Combine this with other filtration methods for fully clean water.

Reverse Osmosis: Use reverse osmosis systems to remove contaminants like heavy metals, nitrates, and fluoride from your water using a specialized membrane. Keep the membrane filters in good condition and perform regular maintenance for optimal results.

Boiling: The simplest method? Boil your water! Bring it to a rolling boil for at least one minute (or three minutes if you're at high altitude) to kill bacteria and other pathogens. Let it cool before drinking or using.

FOOD PRODUCTION

Food Production

Gardening & Crop Rotation

As covered in an earlier chapter, gardening is truly the cornerstone of self-sufficiency. It's all about growing a mix of fruits, veggies, and herbs right in your backyard. Let's dive into some key gardening tips for homesteading:

- **Site Selection:** Pick a sunny spot with good, well-drained soil for your garden. It helps if water sources are nearby for easy irrigation.
- **Crop Selection:** Choose crops that are well-suited to your climate and soil. Consider the growing season, water requirements, and nutritional benefits when deciding what to plant.
- **Companion Planting:** Use companion planting techniques to boost plant health and productivity. Pairing certain plants together can improve soil fertility, reduce insect issues, and make better use of space.
- **Crop Rotation:** To keep soil fertile and fend off pests and diseases, rotate your crops. Swap them among different plant families to maximize yields and minimize nutrient loss.

Livestock Rearing for Self-Sufficiency

Do you already grow your own fruits and veggies? Great! How about raising some animals to become even more self-sufficient? Having livestock means a steady supply of protein, dairy, and other goodies. Let's dive into adding livestock to your off-grid homestead:

- **Choosing the Right Livestock:** First things first, pick animals that match your climate and what you have available. Think about space, food supply, and local rules. It all matters when deciding which animals will fit in best.
- **Housing and Infrastructure:** Build comfy homes for your animals so they stay healthy. Put up shelters, fences (don't forget gates), and spots where they can eat safely, away from predators and bad weather.
- **Feeding and Nutrition:** Plan out what your animals will eat so they get all their nutrients. Use a mix of grazing, hay, grains, and extra feed. This helps them grow strong and happy.
- **Looking After Them:** Keep an eye on their health with regular vaccines, deworming, and check-ups. Catch any issues early to prevent big problems later.

HANDLING WASTE AND RECYCLING

Composting Methods

Thinking about giving composting a try? It's a fantastic way to cut down on landfill waste and boost your garden soil with rich, organic matter. Let's dive in and see how you can start:

- **Backyard Composting:** Find a spot in your yard and set up a compost area. You can use old pallets or wire mesh to create a simple bin where you'll toss your compost materials.
- **Green & Brown Matter:** Remember the magic balance of composting? It's all about mixing brown materials (like cardboard, straw, and dead leaves) with green materials (such as fruit and veggie scraps, grass clippings, etc.). Layer these ingredients in your compost bin to help them break down more effectively.
- **Aeration & Moisture:** Make sure to turn the pile now and then with a pitchfork or compost tool. This helps get air into the pile and speeds up decomposition. Also, keep it moist—but not soaked; think of the consistency of a damp sponge.

Greywater Recycling Systems

Got a bunch of wastewater from things like taking a bath, doing laundry, and washing dishes? You can cut down on your water use and help the planet by recycling it. Let's figure out how to set up a greywater recycling system.

- **Greywater Collection:** Get yourself a setup to collect all the wastewater from your sinks, showers, and washing machines. It doesn't have to be fancy—just divert those plumbing pipes into a storage tank. Keep it simple.
- **Filtration and Treatment:** Before you reuse any greywater, it needs some cleaning up. You've got to get rid of contaminants and germs. Use filtration systems like sand filters or even create small wetlands. These setups, like bioremediation systems, clean the water enough so you can use it for watering your plants.
- **Irrigation and Landscape Design:** Here's where your garden gets the good stuff! Guide that cleaned-up greywater to your plants using a special irrigation system. Drip irrigation or soaker hoses work best—they get water right where the plants need it without wasting any through evaporation or runoff.
- **Monitoring and Maintenance:** Check up on your greywater system regularly. You want to make sure everything is running smoothly. Keep an eye on water quality and how the system is performing. Fix things up as needed to keep it effective.

Minimizing Environmental Impact

There are lots of ways to go green on your off-grid homestead. Besides composting and recycling greywater, check out some other cool ideas:

- **Reduce Single-Use Plastics:** Ditch the single-use plastics! Go for reusable stuff like stainless steel water bottles, cloth grocery bags, and glass containers for your food.
- **Energy Conservation:** Save energy by choosing appliances that don't gobble up power. Turn off lights and gadgets when you don't need them. Consider using solar power and wind energy too!
- **Sustainable Transportation:** Cut down your carbon footprint by riding a bike, walking, or sharing rides with friends when you can. For longer trips, think about getting an electric or hybrid car.
- **Minimalist Lifestyle:** Keep things simple. Declutter your space and focus on quality over quantity. Avoid unnecessary shopping sprees, choose items that last, and fix them if they break instead of buying new ones.

STRATEGIES FOR SAFETY AND DEFENSE

Perimeter Defense

Securing the perimeter of your off-grid homestead is key to keeping your place safe and sound. Here's how you can do it:

- **Fencing:** Put up a strong fence around your property to keep unwelcome guests out. Go for materials like chain-link, wood, or metal because they last longer and can take a beating.
- **Security Gates:** Set up security gates where people come in. Use tough locks and sturdy hinges to stop folks who shouldn't be there. Consider adding motion-activated lights so you can see what's happening at night.
- **Surveillance Cameras:** Place surveillance cameras around the edges of your property. They help you monitor activity and might make bad guys think twice. Wireless cameras with night vision are ideal since they work even in the dark.

- **Guard Dogs:** Train some guard dogs to patrol your land and alert you if something seems off. Breeds like German Shepherds or Rottweilers are great—they're loyal, smart, and effective when properly trained.

Home Security Measures

Keeping your off-grid home protected brings peace of mind and safety from potential trouble. Consider these measures:

- **Reinforced Doors and Windows**: Make your doors and windows tougher with heavy-duty locks, deadbolts, and security bars to prevent break-ins. Shatterproof glass is also a smart option for extra protection.
- **Alarm Systems**: Get an alarm system that covers doors, windows, and detects motion or glass breaking. It'll alert you if someone's sneaking around. Linking it to a monitoring service enhances security even more.
- **Safe Room**: Designate a spot in your house as a safe space for emergencies. Reinforce the walls, doors, and windows—using bulletproof materials if possible. Stock it with emergency supplies, communication devices, and first aid kits.
- **Firearms and Self-Defense Training**: Consider firearms for home defense if you're comfortable with them, but ensure you get proper training first. Additionally, sign up for self-defense classes so you know how to protect yourself and your loved ones when needed.

Community Collaboration for Safety

Building strong community collaboration is super important for making your off-grid area safe and secure. Here's how you can team up with your neighbors:

- **Neighborhood Watch Programs:** Start a neighborhood watch program to encourage everyone to look out for each other. Report any strange activity to the local police. Also, have regular meet-ups and training sessions to boost community awareness and preparedness.
- **Communication Networks:** Set up communication networks with nearby homesteads. Share info, resources, and support when needed. Use two-way radios or cell phones to stay connected and handle emergencies together.
- **Mutual Aid Agreements:** Create mutual aid agreements with neighboring communities. This way, you can help each other out during emergencies or crises. Work together on emergency plans, share resources, and come up with defense strategies to make your community stronger.

Health and Medical Preparedness

First Aid Training and Supplies

- **Take a First Aid Course:** Enroll in a certified first aid course. Learn vital skills like CPR, handling fractures, and treating wounds. Online courses are also available if you prefer studying from home.
- **Stock Up on First Aid Supplies:** Assemble a complete first aid kit with essentials like bandages, tweezers, adhesive tape, scissors, and antiseptic wipes. Check and restock your supplies regularly to always be ready.
- **Learn Basic Medical Procedures:** Get familiar with basic medical procedures—suturing wounds, splinting fractures, and giving medications. Practice these skills under the guidance of a trained professional if you can.

NATURAL HEALING TECHNIQUES AND HERBAL TREATMENTS

Besides regular medicine, herbal remedies and natural healing methods are super helpful for staying healthy and fixing small problems. Let's see how to fit them into your plan:

- **Check Out Herbal Remedies:** Look into herbs like echinacea, ginger, and garlic. These have awesome immune-boosting and germ-fighting powers. Consider starting a small garden with these medicinal herbs so you have them on hand.
- **Learn Natural Healing Methods:** Get to know practices like acupressure, aromatherapy, and homeopathy. These can help with pain, stress, and overall well-being. Practice them often to get good at them.
- **Talk to the Experts:** Consult herbalists and naturopaths for advice. They can help create special herbal remedies tailored to your needs.

Long-Term Healthcare Planning

Making plans for the long run ensures you get the care you need in tough times. Here's how to be ready for what's ahead:

- **Set Up Healthcare Directives:** Write down your healthcare wishes in case you're unable to express them later. Choose someone you trust to make decisions for you if needed.

- **Get Health Insurance:** Secure health insurance to help cover medical bills. Consider options for catastrophic insurance as well; it can be a big help in major emergencies.
- **Form a Support Network:** Build a strong network of family, friends, and healthcare professionals who can assist when needed. Keep everyone in communication and working together for the best outcomes.

NETWORKING AND COMMUNICATION LINKS

Establishing Off-Grid Communication Channels

As we discussed in Chapter 13, keeping in touch and staying informed during emergencies is super important. Here's a guide on setting up off-grid communication:

- **Invest in Two-Way Radios:**
  Get yourself some two-way radios with long-range features. They're great for chatting with family and neighbors when you need reliable communication. Pick models that have multiple channels and privacy codes to keep your chats secure.
- **Set Up a Communication Plan:**
  Make a solid communication plan. It should detail how you'll contact loved ones, share important info, and coordinate your actions. Give everyone in your group specific roles and responsibilities.
- **Practice Emergency Communication Drills:**
  Hold regular drills to check if your communication plan works well. Everyone should know what to do. Identify any weak spots or areas that need improvement, then tweak your plan as needed.

Alternative Internet Access Options

Staying connected to the internet is key—for information, contacting emergency services, and keeping in touch with the outside world. Here are some options to consider:

- **Satellite Internet:**
  Think about satellite internet for reliable connections in remote or off-grid locations. You can set up satellite dishes on rooftops or other high spots to get high-speed internet.
- **Mobile Hotspots:**
  Grab a mobile hotspot that uses cellular networks for internet on-the-go. Choose plans that offer sufficient data and coverage for your needs.
- **Mesh Networking:**
  Create a mesh network using devices like smartphones, laptops, and routers. This establishes a decentralized network where you can communicate and share data without needing central infrastructure.

Radio & Ham Radio Operations

Radio communication still stands as a trusty way to talk over long distances, especially useful in far-off or disaster-hit places. Here's how you can dive into ham radio operations:

- **Get a Ham Radio License:** Start by getting a ham radio license from the Federal Communications Commission (FCC) to operate amateur radio gear legally. Study for the licensing test using online guides and practice questions. It's not too hard if you put in some time.
- **Invest in Ham Radio Gear:** Spend some money on ham radio gear like transceivers, antennas, and power supplies. Pick equipment that suits your chatting needs and won't break the bank. Consider frequency bands and transmission power before buying.
- **Join a Ham Radio Group:** Become part of local ham radio clubs or join online groups. It's a great way to meet other enthusiasts, join events, and learn a lot. Regularly participating in radio chat exercises helps sharpen your skills and expand your network.

PROJECT 23: BUILD YOUR OWN GREENHOUSE FOR YEAR-ROUND GARDENING

This project is one of the longest in the book, and it is a lot more like building a home, just on a much smaller scale. You can modify the instructions to make something that will fit within whatever space you have available, whether it is a large yard, a suburban house with more house than yard, or an apartment.

Materials and Preparation

Before you start, you will need the following materials:

- Moisture resistant lumber
- PVC pipes
- Aluminum
- Covering (this can be glass, polycarbonate sheets, or plastic sheeting)
- A greenhouse schematic that uses the foundation, frame, and covering your want for your greenhouse.

There are many variations on greenhouses, so what you need is going to depend on how much space you have and the type of greenhouse you want. If you are in a small space, you don't want a glass greenhouse. These instructions are fairly general, but it highlights all of the necessary steps, regardless of the type of greenhouse you want.

Steps

1. Determine where you will put the greenhouse and what dimensions will fit the available space. As you do this, consider the following
   a. It needs to get enough light for the plants to grow and thrive.
   b. Locate near water as you will need to water your plants regularly.
   c. If possible, have near an electrical source.
2. If you need a foundation, you need to add one. This won't apply in many situations, but if you plan to add the greenhouse to ground that isn't level, you will need to add the foundation before you start the process of building your greenhouse. You will also need a foundation if you want a real structure, which you won't need for a much smaller greenhouse.
3. If you don't need a foundation, add gravel, bricks, or other material so that water can drain through the floor.
4. Build the frame. The following pictures are for a smaller more traditional greenhouse. If you plan to make a PVC greenhouse, the steps are drastically different, as are the steps for a completely glass greenhouse. Follow the greenhouse schematics you chose.
   a. Glass greenhouses should be constructed on flat, raised surfaces. Making them an A frame is the simplest and often yields the best results. You may need to have glass specially cut.
   b. The wood you use should be chosen based on the environment to ensure it will not rot or have issues because of the amount of humidity and water that will be encouraged in it.
   c. PVC greenhouses can be built from kits, and are the simplest of the type. The biggest draw back is the PVC covering may get discolored over the years.

Glass Greenhouse

Making a Wooden Frame

PVC Greenhouse

5. If large enough, install a watering system.
6. If you want to add a power source, connect the greenhouse to the nearby power source.

PROJECT 24: MAKE A HINGED HOOPHOUSE FOR YOUR RAISED GARDEN BED

If you don't want to build a full greenhouse, you can add Hinged Hoophouses to your raised beds. This will help them grow and give you a way to protect your crops and plants.

Materials and Preparation

Before you start, you will need the following materials:

- Polypipe
- Plastic pipe cutters
- Pipe clips
- Tarp
- Cast iron handle
- Hinges
- Drill
- Cabin hook (to protect against high winds)
- Stapler gun
- Screws and washers that are durable in wet weather
- Timber
- Wooden slats

Steps

1. Measure the lumber against the shortest side of your current raised bed. Repeat this for the other side and the back side of the bed.
2. Drill holes into the wood, then screw all three pieces into the bed.

Adding the Frame

3. Cut a support piece for the pipe at the top.
    a. Measure to the top.
    b. Clamp the piece in place.
4. Cut the pipes based on your measurements.
5. Screw a clip in the inside corner along the longest piece. Repeat on the other side.
6. Insert the piping into the clip.

Adding the Piping

7. Screw the pipes in position.
8. Install the supporting slats.
9. Cover the roof with the plastic.

Add the Covering

10. Staple the plastic in place.
11. Add screws to the frame.
12. Tighten the screws for the covering.
13. Drill holes where you want to install the hinges along the base and the lid.
14. Drill on the lid where you want to add the handle.
15. Install the handle where you drilled the holes.

PROJECT 25: DIY STORAGE SHED FROM RECYCLED PALLETS

Wood pallets have many uses outside of moving large, bulky objects, and one of the best is creating a shed for storing tools, equipment, and even firewood.

Materials and Preparation

Before you start, you will need the following materials:

- 9 pallets in good condition and that are the same size (this projects uses pallets that are 48 x 40 in)
- 4 4x4 in x 10 ft posts
- 6 2 x 4 in x 8 ft boards
- 3 1 x 4 in x 10 ft. boards
- 4 2 x 4 in x 8 ft boards
- 2 26 in x 10 ft galvanized roofing metal
- 6 galvanized rafter ties
- 3 in deck screws
- 1.5 in metal roofing screws
- 4 wooden stakes
- Concrete mix
- Wheelbarrow
- Post hold digger
- Shovel
- Saw
- Tape measures
- Metal snips/scissors
- Level
- Step ladder
- Drill
- Hammer
- Framing square

Steps

1. Determine where you want to build the shed. Make sure you don't have any restrictions, such as water and utility systems in your desired locations. You don't want to encounter anything when you set the posts.
2. Create the pallet assembly.
    a. Cut the 1 x 4 in lumber into 1 ft long pieces. There will be 10 when you finish.
    b. Screw two of the joiner pieces on two pallets.
    c. Cut another pallet in half.
    d. Screw the two pieces to the back of the joined pallets.
    e. Cut two other pallets to have a 32 in section.
    f. Screw these along the length of the pallets, making the side walls of the shed.
3. Set the shed's posts.
    a. Level the ground where you plan to make your shed.
    b. Lay the floor, then mark the corners against the ground.
    c. Use the post hole digger to make the holes for the posts. They should be around 9 in in diameter.
    d. Put the 4x4 in posts in the holes, and verify it is level.
    e. Brace the posts and secure them with the cement, then wait 48 hours for it to dry.
4. Add the walls.
    a. Place the floor pallets back between the posts.
    b. Screw the wall pieces to the posts.
    c. Join the wall and back pallets.

Securing the Sides

5. Install the top.
    a. Cut the rest of the lumber to create rafters.
    b. Add the galvanized roof tiles on the rafters.
    c. Screw the treated pieces to the rafters, running perpendicular and down the middle.
6. Add the roof.
    a. Cut the sheet metal panels.
    b. Place the pieces on the rafters.
    c. Screw the pieces in places.

20.Crafting a Survival Kit

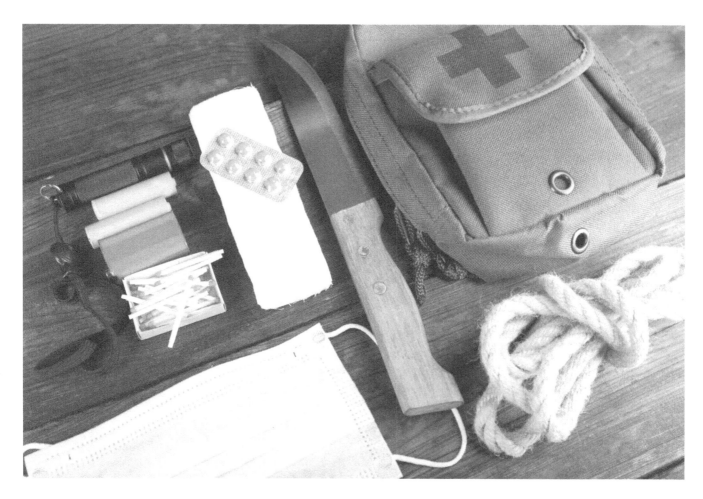

OBJECTIVE OF A BUG OUT BAG

A Bug Out Bag: Your Emergency Lifeline

A Bug Out Bag, simply put, contains the basic stuff you'll need to get through at least 72 hours during a crisis or disaster. It's like your emergency best friend.

Importance of Preparedness

Having a well-stocked Bug Out Bag significantly boosts your chances of staying safe. You'll have the right resources to keep you and your family out of harm's way. Being prepared isn't just about hoarding supplies; it's also about knowing what to do, having certain skills, and maintaining the right attitude to handle tough situations and take care of your loved ones.

Overview of Building a 72-Hour Survival Kit

Creating a 72-hour survival kit is essential for emergency situations. The idea is simple: in major disasters, emergency crews might take up to three days to reach affected areas. That's why your Bug Out Bag should have everything you need to survive on your own for at least three days. Think food, water, shelter, first aid supplies, essential tools, and other must-have items.

AUXILIARY EQUIPMENT AND TOOLS

Here's some crucial stuff you need in your bug-out bag:

- **Multi-Tool:** A multi-tool is like a little hero. It packs in pliers, knives, screwdrivers, and more, all in one spot. You can fix gear or make shelter with it. Truly, it's a must-have for anyone who loves being outdoors.
- **Fixed Blade Knife:** This knife is your go-to buddy for cutting, slicing, and carving. Plus, if things get rough, it can help keep you safe. Look for one with a good grip and a strong blade that handles tough use well.
- **Folding Saw:** Building shelter? Gathering firewood? Making tools? A folding saw is your friend here. Pick one that's small yet strong with sharp teeth to make cutting wood easier. It's great because it's portable and super useful.

Signaling and Navigation Equipment

When you're out in the wild, staying on track and being able to signal for help is key. Here's what you need:

- **Compass:** Your trusty guide. A compass helps you find your way. Learn to use it with a map to figure out direction, make routes, and know where you are no matter where you go.
- **Signal Mirror:** Sometimes words just don't work, but a signal mirror sure does. Use it to flash sunlight towards rescuers—they can see you from miles away! It's small and light; definitely a must-have in emergencies.
- **Whistle:** A loud whistle can be heard from really far away. It alerts others that you're there and signals for help. Keep it handy so you can use it fast in urgent moments.

Lighting & Fire Starting Tools

Fire is a must-have for staying alive. It keeps you warm, gives light, and helps cook your food. Let's see how you can make fire anytime, anywhere:

- **Waterproof Matches:** Keep these matches in a dry, safe place in your backpack. They work even if it's wet outside, so you can always light a fire when you need to.
- **Fire Starter:** Get a good fire starter like a ferrocerium or magnesium striker. It's super helpful to have something reliable. Make sure to practice using it often so you're ready to start fires, even when it's tough.
- **Headlamp:** When it gets dark, a headlamp lights up your way without tying up your hands. Look for one with adjustable brightness settings and long battery life. This way, it'll work well even on long trips.

Water & Food Gear

In the wild, having clean water and food is super important. Here's what you'll need:

- **Water Filtration System:** Streams and rivers can be dirty. Use a water filtration system to make sure your water is safe to drink. Read and follow the instructions so it works well.
- **Collapsible Water Container:** If you're packing light, a collapsible container is great for carrying clean water. Pick one that's tough yet lightweight—it needs to handle rough conditions without being too heavy.
- **Fishing Kit:** Bring along a small fishing kit with hooks, line, and sinkers. It's handy for catching fish from lakes or rivers, giving you some fresh, protein-rich food. Practice basic fishing so you're ready to catch dinner when you're hungry.

EMERGENCY CARE AND MEDICAL ESSENTIALS

Importance of First Aid in Survival Situations

Hey, it's super important to know first aid basics. Here's why:

- **Immediate Help**: First aid lets you give quick help to yourself or others if there's an injury. It stops things from getting worse and can even save lives.
- **Stabilization**: By taking care of injuries and handling symptoms, first aid boosts the chances of staying alive until medical pros can step in.
- **Empowerment**: Knowing first aid gives you confidence. It helps you take charge during emergencies, cuts down on panic, and keeps things calm.

Essential Medical Supplies for a Bug Out Bag

Putting together your Bug Out Bag? Don't forget medical supplies! Here's a quick list of must-haves:

- **First Aid Kit**: Get a basic first aid kit with bandages, gauze pads, sticky tape, antiseptic wipes, tweezers, scissors, and gloves. Keeping it well-packed and organized makes it easy to find what you need fast.
- **Trauma Supplies**: Add items like trauma shears, special dressings to stop bleeding, and tourniquets for treating severe wounds.
- **Medications**: Bring essential meds like painkillers, allergy meds, anti-diarrhea pills, and any prescriptions for you or your group.
- **Personal Protective Equipment**: Gloves, masks, and eye protection are key. They prevent infections and keep things clean when providing first aid.

Tips for Organizing & Accessing First Aid Supplies

- **Container Selection:** Pick a tough, waterproof box to keep your supplies safe from the environment. You don't want anything getting wet or ruined.
- **Labeling:** Label everything! Use clear labels on each compartment in your kit to find items quickly. Color-coding or symbols can help too.
- **Accessibility:** Keep your first aid kit in a spot where you can grab it fast. Consider attaching it to the outside of your bag for easy access.
- **Regular Inspection & Restocking:** Regularly check your first aid kit to ensure everything is present and in good shape. Check expiration dates, and if something's used or expired, replace it right away.

SAFE HAVEN AND SECURITY

Shelter Components for a Bug Out Bag

- **Tarp or Tent:** A lightweight, waterproof tarp or tent will keep you safe from rain, wind, and sun. Get one that's easy to set up and strong enough for rough weather.
- **Sleeping Bag or Blanket:** Stay warm by packing a small but cozy sleeping bag or a wool blanket. It'll help keep the cold away.
- **Ground Pad or Mat:** For comfort and warmth, use a pad or mat under your sleeping bag. An inflatable one or a foam pad works best for insulation and cushioning.

Protective Gear for Weather & Wildlife

Here's the stuff you need when out in nature & braving the elements:

- **Weather Protection:** Make sure to pack waterproof and windproof layers. You'll need a rain jacket, some pants, and a few insulated layers to keep warm and dry when the weather decides not to be your friend.
- **Insect Repellent:** Don't forget insect repellent! It helps keep away those annoying bugs & protects against nasty diseases. Look for one with DEET or picaridin.
- **Wildlife Deterrents:** If you're heading to bear country, bring along some bear spray or other wildlife deterrents. A whistle or air horn can also come in handy for scaring off any curious critters.

Tips for Setting Up Camp and Staying Safe

Setting up camp safely is super important for a good time outdoors. Here are some simple tips:

- **Location Selection:** Pick a spot that's flat and dry. Stay clear of any places where things might fall, flood zones, or wildlife trails.
- **Tent Pitching:** Find a sheltered area to pitch your tent or tarp. Make sure it's anchored well so it doesn't blow away in the wind! Use guy lines & stakes to keep everything stable.
- **Fire Safety:** Keep your campfire at a safe distance from your tent to avoid accidents. Clear away any vegetation around the fire spot and create a fire ring or pit to contain it.
- **Leave No Trace:** Follow Leave No Trace principles. This means minimizing how much you impact the environment. Dispose of waste properly & make sure not to disturb any wildlife homes.

ADDITIONAL ITEMS AND CONSIDERATIONS

Miscellaneous Items & Considerations

Extra Essentials for Complete Preparedness

If you have the room, think about adding these to your Bug Out Bag for a well-rounded kit:

- **Duct Tape**: This is the ultimate fix-all. It can repair gear, patch shelters, and even work as a makeshift bandage in a tough spot. Keep a roll handy for unexpected repairs.
- **Paracord**: Light but super strong, paracord has countless uses in survival situations. You can build shelters or make slings with it. Having some on hand is a game-changer.
- **Emergency Cash**: If there's a long-term emergency, accessing funds might be tricky. Stash some cash in your bag. You'll need it to buy supplies or get a ride when needed.

Navigation Aids & Communication Devices

To navigate and communicate effectively, here's what you need:

- **Topographic Map**: A detailed topographic map of your area is key for finding your way through unknown places. Familiarize yourself with the main landmarks and features to pick the safest and quickest routes.
- **Two-Way Radio**: When cell service is down or unavailable, a two-way radio lets you communicate with your group or emergency teams reliably. Choose one with good battery life and enough range for your needs.
- **GPS Device**: For precise navigation and tracking, a GPS device is crucial. Look for one with preloaded maps, waypoint marking, and a durable design suitable for outdoor use.

Personal Hygiene & Comfort Items

Keeping up with personal hygiene and comfort can really lift your spirits during an emergency. Here's what you need to stay fresh & comfy:

- **Travel-sized Toiletries**: Make sure to pack key toiletries like toothpaste, toothbrush, soap, and hand sanitizer. These help you stay clean out there. Pick travel-sized bottles—they save space and weight in your bag.
- **Quick-Dry Towel**: Grab a compact, quick-drying towel. It's lightweight and perfect for staying clean and dry while on the move. Look for one made from microfiber or something similar that soaks up moisture quickly and dries fast.
- **Comfort Items**: When things get tough, having comfort items like a favorite book, journal, or family photo can really make a difference. They give emotional support and feel like a touch of normal life. Toss a few of these small items into your bag for a morale boost during hard times.

ENHANCING EFFICIENCY AND PLANNING AHEAD

Strategies for Reducing Weight & Maximizing Functionality

As you're working on refining your Out Bag setup, here are some friendly tips for streamlining weight and maximizing functionality:

- **Prioritize Essentials:** Pack only the must-have items that can serve many purposes. Think about what will help you survive and be comfortable. Each thing you pack should be useful in more than one way. Get rid of anything bulky or unnecessary.
- **Trim Excess Gear:** Take a good look at your gear often. See where you can cut down on weight by choosing smaller, lighter gadgets. Multi-use tools can be a great way to save space while still getting the job done right.
- **Keep an Inventory:** Know what's in your bag. Keep track of the items and make sure nothing is expired. This means no carrying food, medicine, or tools that you can't use anymore.
- **Practice Loadouts:** Do practice runs with your bag to see how well it works in real life. Try different setups and see which one feels best. Make tweaks as needed for the best performance.

21. Emergency Medical Care

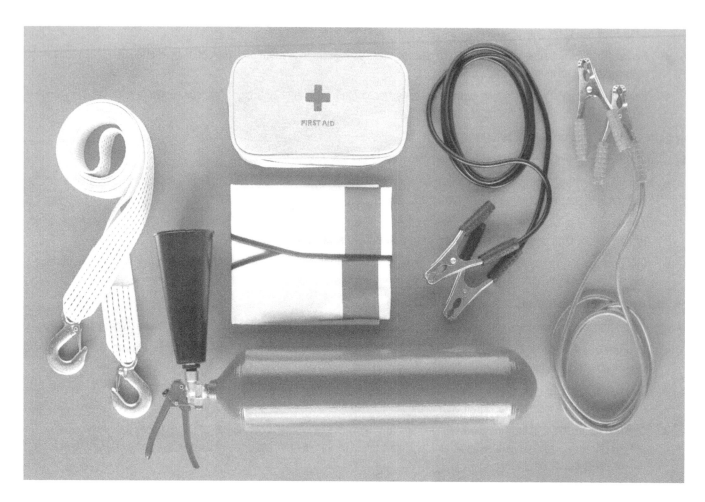

ASSEMBLING YOUR WILDERNESS FIRST AID KIT

Heading out to the wilderness? Your first aid kit is your go-to medical buddy. It's super important to pack right—think handy & compact but still super useful. Here's a guide on how to put together your wilderness first aid kit:

- **Essential Medical Supplies:** Fill your kit with a good variety of things. Don't forget these:
  - **Bandages and Dressings:** You'll need stuff for wound care. Get adhesive tape, gauze pads, and bandages in different sizes.
  - **Antiseptic Solutions:** Pack alcohol wipes, iodine swabs, or antiseptic sprays. They help clean wounds & stop infections.
  - **Medications:** Bring pain relievers and antihistamines for allergies, plus any personal prescriptions you might need.
  - **Splinting Materials:** Grab SAM splints, triangular bandages, or even rolled-up magazines. These help stabilize fractures.
  - **Thermometer:** You gotta keep an eye on body temperature. It helps spot hypothermia or fever.
  - **Tweezers and Scissors:** Handy for little tasks like cutting bandages and pulling out splinters.
  - **Emergency Blanket:** Reflective blankets are great! They keep you warm and can give shelter in tight spots.
  - **CPR Mask:** This little tool lets you safely give CPR if needed (hope you never have to).

- **Customization for Specific Environments:** Make your kit fit your surroundings:
    - **Climate:** Think about temperature extremes, humidity, & rain when picking meds and dressings.
    - **Terrain:** Pack extra stuff for common injuries in the area like sprains from rocky places or cuts from sharp plants.
    - **Duration of Trip:** Make sure you have enough supplies for however long you're out there. It's better to have a little more than run out too soon.

EVALUATING MEDICAL EMERGENCIES AND FIRST RESPONSE ACTIONS

Performing a Primary Assessment in the Wilderness

Performing a primary assessment in the wilderness requires a quick and accurate evaluation of your surroundings. Here's how to break down this vital task into easy steps:

- **Size Up the Scene:** First, take a moment to look around. Make sure both you and the patient are safe. Watch out for dangers like falling trees, wild animals, or rough ground.
- **Approach with Caution:** Stay calm as you approach the patient. Reassure them by saying help is coming. Look for signs of pain or stress.
- **Check Responsiveness:** See if the patient is awake. Gently tap them and ask, "Are you okay?" Watch for any response, like moving, moaning, or speaking.
- **Airway, Breathing, Circulation:** If the patient is conscious, focus on the ABCs – essential for life-threatening emergencies.
    - **Airway:** Tilt their head back slightly and raise the chin to see if the airway is clear. Check breathing by looking, listening, and feeling for no more than 10 seconds. Use alternative techniques if a spinal injury is suspected.
    - **Breathing:** Observe their chest movement. Listen for unusual sounds like gasping or wheezing. Start rescue breaths if there's no breathing.
    - **Circulation:** Feel for a pulse at their neck (carotid artery). If there's no pulse, start CPR immediately.
- **Address Immediate Threats:** Be alert for severe bleeding, spinal injuries, or cardiac arrest. Treat these issues quickly to stabilize the patient.

In the wilderness, resources might be scarce and getting help can take time. Focus on what's most critical with the resources you have available.

CARING FOR INJURIES IN THE WILD

When you're out in the wild and need to treat a wound, you often have to think on your feet. Here's a simple guide to cleaning and dressing wounds with whatever is handy:

- **Stop the Bleeding:** First things first, you gotta stop the bleeding! Use a sterile dressing or a clean towel. Press directly on the wound. If possible, lift the injured limb above the heart. This helps slow down the bleeding.
- **Cleanse the Wound:** Once the bleeding is under control, it's time to clean it up. Wash the area with some saline solution or clean water to get rid of any bacteria and dirt. Avoid using stuff like hydrogen peroxide or alcohol—they can hurt tissue.
- **Dress the Wound:** Next, cover the wound with a bandage or dressing. If you don't have those, no worries. You can use clean cloth, gauze, or even pieces of clothing. Hold everything in place with a temporary bandage or some adhesive tape.
- **Monitor for Infection:** Keep an eye on the wound for any signs of infection like redness, swelling, warmth, or pus. If any of these show up, get medical help ASAP.

Different kinds of wounds need special care:

- **Punctures:** For puncture wounds (like from nails or animal bites), make sure to clean the area well and put on some antibiotic ointment if you have it. Watch out for signs of infection since these can lead to deep tissue damage.
- **Lacerations:** Cuts (aka lacerations) should be cleaned and closed up with adhesive strips or butterfly bandages if the edges come together easily without stress. For deep or jagged cuts, see a doctor because you might need stitches.
- **Abrasions:** For scrapes (abrasions), give them a gentle cleanse. Then add some antibiotic ointment and cover with a clean bandage or dressing to help it heal.

METHODS FOR SPLINTING AND IMMOBILIZATION

Creating Effective Splints

Out in the wild, getting hurt is a real possibility. Fractures and dislocations can happen, so it's important to know how to make a good splint with what you've got around.

- **Look for strong materials:** Branches, trekking poles, or even rolled-up clothes can do the trick. Once you've gathered your supplies, assess the injured limb and the area around it to determine the best way to immobilize it.
- **Emphasizing stability and comfort:** Keeping everything stable is crucial. You don't want to make things worse while trying to move. Just be careful not to make it too tight; you don't want to cut off blood flow. Also, consider comfort by using soft materials like clothing or bandages to pad the splint. This will prevent rubbing and discomfort.
- **Techniques for immobilizing fractures and dislocations:** For fractures, gently align the injured limb before applying the splint. Use triangular bandages or strips of cloth to secure everything in place, ensuring the splint covers both the injured area and the joints above and below it.
- **Dislocations:** If it's a dislocation, wrap the joint with a splint to provide support and prevent further movement.

HANDLING BITES AND STINGS

Identifying Common Bites & Stings:

Out in the wild, bites from insects and animals happen all the time. Mosquitoes, ticks, snakes, and spiders are just a few culprits. Know the signs: redness, swelling, pain, and even numbness in the affected area.

Treatment Guidelines:

Act fast when you get bitten or stung to prevent things from getting worse:

- Clean the spot with soap and water to remove venom or bacteria.
- Use a cold compress to help with pain and swelling (it really works).
- If possible, elevate the limb; it slows down the spread of venom.

Minimizing Risk and Addressing Complications:

Prevention is crucial to stop bites and stings before they happen:

- Wear protective clothing and use insect repellent to keep bugs away and reduce risk.
- If an allergic reaction occurs, administer antihistamines or epinephrine immediately and seek medical help fast.

Strategies for Wilderness Survival:

Stay alert out there:

- Avoid areas where dangerous critters hang out—especially at night when they're more active.
- Always carry a first aid kit ready for bites and stings. Stock it with essentials like antiseptic wipes, tweezers, and adhesive bandages.

MANAGING THERMAL INJURIES: BURNS AND FROSTBITE

Burns & frostbite are common injuries in the wild. They happen due to fire, chemicals, or extreme cold. Knowing how to act quickly and properly can save the day.

Immediate Care Protocols for Burns:

- **Assess the Severity:** Check how big and deep the burn is, and see where it is on the body.
- **Remove the Source:** If fire or chemicals caused the burn, get the victim away from it quickly.
- **Cool the Burn:** Rinse the burn with cool, running water for about 10 to 20 minutes. Don't use ice—it can make things worse.
- **Protect the Wound:** Cover the burn with a clean, dry dressing to prevent infection. Only use ointments or creams if a doctor tells you to.

- **Seek Medical Attention:** For severe burns, get help fast. Use any communication devices available to call for assistance if you're in the wilderness.

Immediate Care Protocols for Frostbite:
- **Recognize the Symptoms:** Check for signs like numbness, tingling, discoloration (white, blue, or gray skin), and skin that feels hard or waxy.
- **Protect from Further Cold Exposure:** Move the person to a warm place. Take off any wet clothes. Don't rub or massage—this can cause more harm.
- **Gradual Rewarming:** Put the frostbitten area in warm (not hot) water, between 37-40°C (98.6-104°F), for 20-30 minutes. If there's no water, use body heat.
- **Handle with Care:** Don't rewarm if there's a chance of freezing again before getting medical help.
- **Seek Medical Attention:** Frostbite needs professional care. If medical help isn't close by, consider moving the person to a place where they can get it.

MANAGING ILLNESS IN REMOTE LOCATIONS

When you head out into the wild remote places, you gotta be ready for any sickness that might come up. Spotting symptoms early on and knowing how to handle common health issues can really help in taking good care of yourself.

Wilderness Illnesses & Symptoms
- **Gastrointestinal Issues:** Diarrhea, vomiting, and stomach aches might point to food poisoning, contaminated water, or infections in your gut.
- **Respiratory Infections:** Got a cough, fever, stuffy nose, or trouble breathing? These could be signs of lung problems like pneumonia or bronchitis.
- **Vector-Borne Diseases:** Watch out for fever, skin rashes, achy joints, and extreme fatigue. These could indicate diseases spread by ticks, mosquitoes, or other critters.
- **Dehydration:** Feeling extremely thirsty, dry mouth, dark urine, fatigue, and dizziness? That's dehydration creeping up on you. It can strike fast, especially in hot and dry conditions.

Guidelines for Handling Chronic Conditions
- **Medication Management:** Pack all necessary medications in waterproof containers where they're easy to access. Stick to the prescribed doses and schedules even when you're out in the wild.
- **Symptom Monitoring:** Keep an eye on symptoms of chronic conditions and be ready to adjust your treatment plans if needed. Have important medical records and emergency contact information with you.
- **Environmental Considerations:** Be cautious when managing chronic conditions in challenging wilderness environments, such as extreme heat, high altitudes, or rough terrains.
- **Emergency Response Plan:** Have a plan in place for what to do if your chronic condition worsens or complications arise while you're in the wilderness. This includes how to communicate and what steps to take if evacuation becomes necessary.

NATURAL REMEDIES TO PAIN RELIEF

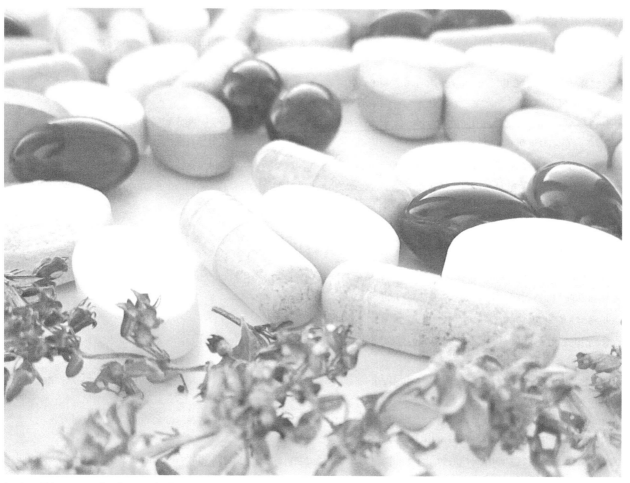

In the wild, you need to know what's out there for pain relief and healing. Let's dive into natural remedies and see their benefits!

Benefits of Natural Remedies:

    Super abundant in nature.

    Few, if any, side effects compared to store-bought drugs.

    Some also have antimicrobial *and* anti-inflammatory perks.

Limitations of Natural Remedies:

- Effectiveness varies depending on you and your condition.
- Not much solid science backing them up.
- Can be toxic or cause bad reactions if misused.

When you head out on wilderness trips, wild plants can really help ease pain and assist your body in healing. Here's a list of plant remedies you might find:

- **Willow Bark**: Contains salicin, which is like aspirin, perfect for pain relief.
- **Arnica**: Good for bruises, sprains, and sore muscles when applied to the skin.
- **Comfrey**: Known for reducing swelling and aiding in wound healing.
- **Chamomile**: Soothes your stomach and helps you relax.
- **Aloe Vera**: Great for burns, sunburns, and skin irritations. It's very soothing.
- **Echinacea**: Boosts your immune system, helping you fight off infections like colds and flu.
- **Lavender**: Calming and antiseptic. Useful for helping you relax and treating minor burns or insect bites.
- **Plantain**: Reduces itching and swelling from insect bites, stings, and small wounds.

- **St. John's Wort**: Used for mild nerve pain and burns. Also known for lifting moods.
- **Yarrow**: Helps stop bleeding and promotes clotting, making it great for small cuts and scrapes.
- **Calendula**: Soothes skin irritation and promotes wound healing. Plus, it's anti-inflammatory and antimicrobial.
- **Ginger**: Relieves nausea, motion sickness, and digestive issues when eaten or made into tea.
- **Peppermint**: Eases stomach problems, headaches, and muscle tension. Use it in tea or on the skin.
- **Thyme**: Helps with respiratory infections and soothes sore throats because it's antimicrobial and antiseptic.

Safety Precautions for Using Natural Remedies
- **Identify Correctly:** First things first, make sure you correctly identify the remedy. Misidentifying can lead to dangerous mistakes.
- **Stick to the Dosage:** Always follow the recommended amounts. Overdose or bad reactions can happen if you don't.
- **Ask for Help:** Talk to someone who knows about these remedies or check reliable sources before trying something new.
- **Watch for Allergies:** If you're allergic to plants, be extra careful. Reactions can be serious.
- **Pregnancy Matters:** If you're pregnant or breastfeeding, consult your healthcare provider before using any natural remedies.

In summary, remember these remedies work alongside conventional medicines, not instead of them. When you're unsure or facing serious health issues, always seek professional medical advice.

MANAGING HYDRATION AND WATERBORNE DISEASES

Water is the elixir of life. It promises us health but can bring illness when it's contaminated. Let's look at waterborne illnesses and how to manage hydration when you're out in the wilderness.

Preventing & Treating Waterborne Illnesses

Symptoms of Waterborne Illnesses:
- Stomach problems like cramps, vomiting, and diarrhea.
- Flu-like symptoms – extreme fatigue, headaches, and fever.
- Dehydration caused by these symptoms.

Treatment Options:
- Drink clean water and use electrolyte drinks to stay hydrated.
- Rest frequently and take it easy if you're experiencing stomach issues.
- If necessary and safe, take antibiotics for certain bacterial infections.

To avoid getting sick from water:
- **Pick Your Water Source Wisely:** Choose clear, running streams or rivers far from potential contaminants.
- **Clean That Water:** Use portable filters, purifying tablets, or boil the water to kill harmful organisms.
- **Store Properly:** Keep water in clean bottles and refill regularly to prevent bacteria growth.

Hydration Management in the Wilderness
- **Knowing the Signs:** Thirst, dark pee, dry mouth. You might also feel tired and dizzy.
- **Smart Hydration Techniques:**
    - Drink water often. Don't wait until you're thirsty to drink; it helps keep you hydrated.
    - Watch the color of your pee. If it's pale yellow, you're doing good.
    - Drinks with electrolytes are great because they replenish the salts you lose from sweating.

When you're out in nature, getting clean water is super important for staying alive. Keeping an eye on your hydration can't be ignored. Knowing the dangers of bad water and using good hydration practices means you'll stay healthy while exploring the wild.

22. Vehicle Maintenance

Vehicle upkeep & repair is super important because you want to be able to rely on your car, especially if there's an emergency. A car is handy for getting you, your family, and all your essentials where they need to be. This part of the guide will cover key tips on keeping your vehicle in tip-top shape. You'll get the know-how you need to keep it running smoothly. Plus, it's just smart to take care of your car for everyday stuff too.

ESSENTIAL VEHICLE MAINTENANCE TASKS

- **Check Fluids and Add When Needed:**
  Keeping an eye on your car's fluid levels is a must. It keeps everything running well and helps your vehicle last longer.
    - **Engine oil:** Grab the dipstick and check the oil level. Make sure it falls within the recommended range. If it's low, add more using the right kind of oil.
    - **Coolant:** Peek into the coolant reservoir. It should be at the right level. Use a mix of coolant & water as suggested in your car's manual.
    - **Brake fluid:** Look at the brake fluid level in the master cylinder reservoir. If it's low, top it up with the correct type of brake fluid.

Conducting Tire Inspections, Rotations, & Pressure Checks

Looking after your tires is really important for keeping safe and making them last longer, especially in emergencies when you need good traction (grip).

- **Tire Pressure:** Grab a tire pressure gauge and check the air pressure in every tire, including the spare. Adjust the pressure as needed to match what the manufacturer recommends.
- **Tire Tread:** Check the tread depth using a tread depth gauge or the penny test. If your tires are worn down below the recommended level, it's time to replace them.
- **Tire Rotation:** To ensure your tires wear evenly and last longer, rotate them regularly. Check your car's handbook for the suggested rotation schedule.

Battery Maintenance Tips

Keeping your vehicle's battery in good shape helps it run well.

- **Visual Inspection:** Check the battery terminals and cables for any signs of corrosion or loose connections. Clean and tighten if needed to maintain a good electrical connection.
- **Battery Terminals:** Use a wire brush or a battery terminal cleaner to scrape off any corrosion from terminals and cable connections. Apply a bit of dielectric grease to prevent future corrosion.
- **Charging:** If you don't have a maintenance-free battery, check each cell's electrolyte level and top up with distilled water if needed. Use a battery charger to keep the battery full when not in use.

URGENT REPAIRS AND ON-THE-ROAD ASSISTANCE

From unexpected flat tires to engine troubles, having the ability to repair your car on the road is crucial. Let's look at what you need to handle emergency repairs and access roadside assistance.

Guidance on Handling Common Vehicle Problems on the Road:

- **Flat Tires:** If you get a flat, stay calm. Here's what you do:
    - **Find a Safe Spot:** Stop your car far away from any traffic.
    - **Engage Hazard Lights:** Let other drivers know you're in trouble.
    - **Assess the Damage:** Check why the tire is flat and look for any holes or damage.
    - **Change the Tire:** Got a spare tire and tools? Follow your car's manual and replace that flat.
- **Engine Overheating:** If your engine starts to overheat, act fast to avoid damage:
    - **Turn Off the Engine:** Pull over safely and turn off your car to let it cool.
    - **Open the Hood:** Wait until it's safe, then open the hood to let out heat.
    - **Check Coolant Levels:** If coolant is low, let the engine cool first before adding more coolant.

Procedures for Accessing Roadside Assistance & Emergency Services:

- **Emergency Contact Information:** Make sure you have a list of emergency contact numbers in your car. This should include roadside assistance services and local authorities. Keep them handy.
- **Call for Help:** If you face a breakdown or emergency, grab your mobile phone and give them a call.
- **Provide Details:** When you talk to the dispatcher, be as clear as possible. Share details about where you are, what's going wrong, and anything else important.

Quick Fixes for Temporary Repairs in Emergency Situations:

- **Tire Repair Kit:** Always have a tire repair kit on hand. These usually come with a plug and reamer for quick fixes on punctured tires.
- **Duct Tape:** You'd be surprised how useful duct tape is! It can help secure loose parts or seal small leaks temporarily.
- **Emergency Fuel:** Sometimes we run out of fuel. Keep a spare fuel canister with gasoline in your vehicle. It can help you reach the nearest gas station without much hassle.

CRUCIAL EQUIPMENT AND REPLACEMENT PARTS

Here's a list of tools and spare parts to keep in your car for those unexpected moments:

Overview of Must-Have Tools:

- **Jack and lug wrench** - Super important for changing tires.
- **Tire Pressure Gauge** - Keep track of tire pressure to avoid blowouts. It's a lifesaver.
- **Multi-Tool or Swiss Army Knife** - Handy for many tasks, like cutting and screwing.
- **Jumper Cables** - Perfect for jump-starting a dead battery with the help of another car.
- **Flashlight** - Shine light in dark spaces under the hood or around your car.
- **Basic Tool Kit** - Include screwdrivers, pliers, wrenches, and other necessary tools for small repairs.

How to Assemble a Comprehensive Roadside Emergency Kit for Your Vehicle:

- **First Aid Kit** - Pack essentials like painkillers, bandages, antiseptic wipes.
- **Emergency Blanket** - Stay warm if you break down in cold weather. It's cozy too!
- **Non-perishable snacks & bottled water** - Stay fueled and hydrated during unexpected delays.
- **Reflective Triangle or Flares** - Alert other drivers of your presence in roadside emergencies. Safety first!
- **Fire Extinguisher** - Be ready to handle small fires under the hood or anywhere else.

VEHICLE PREPAREDNESS TACTICS

Importance of Communication Devices and Emergency Kits

Staying connected is super important, even when you're on the go. Here's why:

- Having communication devices like two-way radios, satellite phones, or even just your trusty cell phone can really save the day during an emergency. They let you call for help and coordinate with others if needed.
- Emergency kits are also a must-have. Stock them with basic supplies like first aid items, water, food, blankets, and some handy tools. These kits offer comfort and essential items during unexpected breakdowns or natural disasters.
  Here are some tips for making your emergency kit rock:
    - Add items that match your needs. Think about any medications you take, spare batteries, and hygiene products.
    - Check the supplies often. Make sure nothing's expired or running low.

Route Planning and Navigation During Evacuation Scenarios

Planning ahead is key! Here are some cool tips for planning routes and navigation:

- Think of multiple evacuation routes before you need them. Consider road conditions, traffic, and any hazards.
- Use GPS systems or offline maps to find alternative ways if roads get busy or tricky.
- Learn about landmarks and waypoints along your chosen paths. It helps a lot if GPS fails.
- Keep up with emergency alerts and road closures via radio or smartphone apps. Adjust your route to stay safe and efficient!

Ensuring Your Vehicle is Ready for Any Emergency Situation

Here's what you need to do so your car's always good to go:

- **Keep up with regular maintenance checks:** These help you spot and fix problems early on, before they turn big. Make sure you're routinely checking things like your car's fluids, brakes, tires, lights, and electrical parts.
- **Create a simple maintenance schedule:** Jot it down somewhere easy to see. List out tasks like changing the oil, replacing filters, and checking the brakes. Stick to what the manufacturer recommends for the best performance.
- **Practice safe driving habits:** This helps reduce wear and tear and lowers the chance of accidents. So remember: no aggressive driving! Keep a safe distance from other cars, stick to speed limits, and follow all traffic rules.

ROUTINE VEHICLE INSPECTIONS AND PREVENTATIVE MAINTENANCE

Importance of Regular Vehicle Inspections & Preventive Maintenance Tasks

- Keeping your car in great shape means staying on top of its care. Here's what to keep in mind:
    - Regular check-ups can spot problems early, saving money on big repairs later. Trust me, it's worth it.

- Doing things like changing fluids, replacing filters, and checking belts keeps your car running smoothly for a long time.

Creating a Maintenance Schedule
- Set up a routine based on your car's manual and what's recommended.
  - Include tasks like oil changes, tire rotations, brake checks, and fluid level checks. Do this regularly.
  - Keep a log of all the tasks you've done and when the next ones are due. This way, you won't miss anything important.

Safe Driving Practices to Extend the Life of Your Car & Enhance Safety
- How you drive matters for your car's health and safety. Here are some tips:
  - Keep a safe distance from other cars so you have time to react if something happens suddenly.
  - Don't drive aggressively—no speeding or rapid starts and stops. It wears out your car faster.
  - Stick to speed limits and follow road signs. Pay attention to weather conditions to stay safe.
  - Drive defensively. Always be aware of what's around you and ready for anything unexpected. This helps avoid accidents and keeps you safe.

PROJECT 26: PROTECT YOUR CAR AND GENERATOR WITH DIY EMP CLOTH

While EMP isn't a particularly high risk problem, being prepared means being ready even for the problems that are less likely. You don't want to be unprepared for a scenario that was always a possibility, even if it was a low one. And if an EMP is used near your home, you don't want it affecting your car or generator. Getting a proper

Materials and Preparation

All you need for this is enough EMP cloth to cover your generators and vehicles that you want to protect if there is an EMP event.

Steps

1. Spread out enough material to cover the object you want to cover. Starting with something small, like a computer, can give you a chance to figure out how best to wrap items with the material.
2. Completely wrap the item with the material, ensuring that there is a little extra. You don't want to leave any part of it exposed as that will negate all of your effort.
3. Continue to wrap other items, particularly your generator and vehicles. Other items that you can wrap are your walkie-talkie and radio.

EMP Protective Cover Material

PROJECT 27: FUEL STORAGE MADE EASY: BUILD AND PRESERVE IT

There are several types of fuel and a number of different types of storage containers. You are probably already familiar with most of them, but this project ensures that you have what you need in the event of an emergency.

Materials and Preparation

Before you start, you will need the following materials:

- Storage containers
- Fuel stabilizer
- Fuel

What you need entirely depends on what kind of fuels you store and how much you plan to store. You can store them underground, or in typical barrels. You will probably want to have a container that you can easily carry as well since you may need to move small amounts of fuel.

Steps

1. Ensure that the storage container or containers are all at least 50 ft from any type of ignition source.
2. Determine what containers you need. They are color coded so that you can tell just by looking at the color of the container. Make sure you follow this color scheme so that you never use the wrong fuel and potentially destroy whatever you are trying fuel.
    a. Blue – kerosene
    b. Green – oil
    c. Red – gas
    d. Yellow – diesel
3. Determine how much fuel you need to store. This will require that you inventory everything powered by fuel. Make a checklist so that you can determine how much you need based on what kind of fuel and how much each of your vehicles or devices require fuel.
4. Once you have the fuel, make sure to properly preserve it. Set a rotation period for. your fuels.
    a. Gasoline starts breaking down in less than a month (30 day).
    b. Diesel can be easily contaminated by moisture and algae when it is stored too long, so it usually only lasts a year.
5. Use a fuel stabilizer when you use a jerry can. Make sure you get the right stabilizer for the fuel.

Different Types of Fuel Containers

23. Psychological Readiness for Off-Grid Survivalists

COMPREHENDING MENTAL READINESS

There's a clear psychological dimension when it comes to being prepared. Getting ready dives into the mental and emotional parts of survival that boost our chances of success. Our mindset and emotional toughness are super important and play a big role in how well we handle crises.

Overview of the mental and emotional aspects of survival:

Survival means having both physical stamina and mental strength. When a crisis hits, our thoughts and emotions can either help or hurt our response. Understanding how stress, fear, and uncertainty affect our minds and actions is key to being mentally prepared. By acknowledging the psychological hurdles we might face during emergencies, we can improve our ability to tackle them.

Importance of psychological resilience in crisis situations:

Psychological resilience is about bouncing back from tough times. It means keeping a positive outlook and staying calm—even when things get scary and uncertain. Being psychologically resilient helps us think clearly, make smart choices, and act decisively when it really matters.

STRATEGIES FOR RESPONDING TO EMERGENCY SITUATIONS

In a crazy emergency, your mind can freak out from stress, fear, and all that stuff. But guess what? There are some tricks to help you handle it:

Deep Breathing: Breathing deeply can make your body chill out. It helps calm your nerves and reduces stress signs. To do it, find a quiet spot, sit comfortably, and breathe in through your nose, making your belly pop out like a balloon. Hold it for a bit, then let it out slowly through your mouth. Feel the tension go away with each breath out. Do this a few times until you feel better.

Visualization Tricks: Another way to cope is by using visualization. Imagine good outcomes and mentally practice doing well in tough situations. This builds confidence against fear and worry. Close your eyes and see yourself handling the emergency calmly and smartly. Picture each step to stay safe and okay, like it's going perfectly smooth. See yourself beating challenges and coming out stronger and fine. Let these confident feelings settle in, knowing you can face whatever comes.

Be Here Now: During an emergency, it's easy to get lost in thoughts about what might happen next or what went wrong before. Stay calm by focusing on now—this is called mindfulness. Pay attention to what you see, hear, and feel right around you. Put your all into whatever you're doing right now, whether it's finding shelter, giving first aid, or talking with others.

Stay in the Know & Have a Plan: Feeling more in control comes from staying updated on what's happening and having a solid plan. Keep up with reliable info through emergency broadcasts, official websites, and community alerts. Talk with trusted friends, family, or other preppers to share info and work together if needed. And hey, make sure you've got a good emergency plan with clear steps for different situations. Go over it often so it's like second nature when something happens.

STRENGTHENING MENTAL FORTITUDE

Cultivating a positive mindset requires building mental resilience that can withstand the toughest challenges. In this section, we'll discuss why optimism and having a realistic outlook are crucial during hard times. We'll also share practical tips for boosting mental resilience and emotional well-being.

Importance of Optimism and a Realistic Outlook:

When we hit rough patches, it's natural to feel overwhelmed. But staying positive? It can change everything. Optimism allows us to find the silver linings even in the darkest of clouds. A realistic outlook, however, helps us see things clearly and make smart choices. The next time you're up against a challenge, remember: A positive mindset, mixed with realism, is your best ally.

Practices for Fostering Mental Resilience

Building mental resilience is like strengthening a muscle—it requires regular practice and dedication. Here are some tips to boost your positive mindset and improve your mental well-being:

1. **Positive Affirmations:** Start your day with kind words about yourself. Remember what you're good at, your goals, and your wins. Say things like "I am strong," "I am resilient," and "I can handle anything."
2. **Gratitude Journaling:** Every day, take a moment to jot down what you're thankful for. It could be a beautiful sunset, a friend's kindness, or simply having food on the table. This helps shift focus from negative thoughts to positive ones.
3. **Mindfulness Meditation:** To stay present and reduce stress, try meditating. Find a quiet spot, sit comfortably, and focus on your breathing. If your mind wanders (which it might), gently steer it back to your breath.
4. **Visualize Success:** Use the power of imagination to see yourself overcoming obstacles and reaching goals. Close your eyes and picture your success. This can boost both motivation and confidence.
5. **Healthy Lifestyle Habits:** Physical health supports mental health too! Prioritize sleep, eat well, and exercise regularly. Mind and body work together.
6. **Seek Support:** Don't hesitate to reach out for help from friends, family, or professionals when you need it. Talking about worries can provide perspective and validation.
7. **Practice Self-Compassion:** Be gentle with yourself, especially during tough times. Recognize your progress and forgive yourself for mistakes or setbacks.

BUILDING MENTAL RESILIENCE

- Self-Care Practices for Promoting Psychological Health
- **Practice Mindfulness Meditation:** Find a quiet spot. Sit comfortably or lie down, and close your eyes. Breathe deeply. how the air moves in and out of your body. Let thoughts come & go without judging them. If you get distracted, just return to your breathing. Practice mindfulness for just five to ten minutes daily—even a little can help reduce stress and sharpen your mind.
- **Engage in Regular Physical Activity:** Make fitness part of your daily life—walking, jogging, cycling, or yoga. Working out can lessen anxiety & ward off sadness since it releases feel-good endorphins. Aim for at least 30 minutes of moderate exercise most days. It will boost your overall well-being.
- **Prioritize Adequate Sleep:** Create a relaxing evening routine and shoot for seven to nine hours of good sleep each night. Make a comfy sleeping space, keep a regular sleep schedule, and cut down on screen time before bed. Good sleep is key for emotional stability, clear thinking, and bouncing back from hard times.
- **Practice Healthy Eating Habits:** Fuel your body right to keep your mind sharp & energy steady. Eat whole grains, fruits, veggies, lean meats & healthy fats. Drink water throughout the day to stay hydrated. Avoid too much junk food, alcohol & caffeine as they can mess with your mood and clarity.

The Impact of Social Support and Problem-Solving Abilities on Developing Resilience:

- **Cultivate Supportive Relationships:** Spend time with friends, family, and neighbors who support you. They will understand and lend a hand when things get tough. Share your thoughts and feelings with those you trust, and be ready to help others too. Strong social ties give you a sense of belonging and security, making it easier to handle stress.
- **Seek Professional Help When Needed:** Don't be afraid to contact mental health professionals if you feel anxious, depressed, or face other mental challenges. Psychologists, therapists, and counselors can provide advice and support. They can help you work through emotions and develop coping strategies.
- **Develop Problem-Solving Skills:** Face challenges head-on and look for solutions instead of getting stuck. Break complex tasks into smaller steps, and come up with ways to overcome obstacles. Be flexible; failures are just chances to learn and grow.
- **Utilize Positive Affirmations:** Use affirmations as part of your daily routine to harness positive thinking. Repeat statements like "I am resilient," "I can overcome obstacles," and "I am surrounded by love and support." This boosts your self-confidence and belief in handling life's challenges.

MENTAL HEALTH FIRST AID STRATEGIES

Overview of Emotional Self-Regulation Strategies for Preppers

Emotional self-regulation means managing your emotions so you stay calm and make smart choices. Here's a breakdown of some practical ways to master it:

1. **Recognize Your Triggers:** The first step is figuring out what makes you feel strong emotions. Really take a look at situations, thoughts, or events that often get you upset.
2. **Practice Mindfulness:** When stress hits, mindfulness tricks like meditation or deep breaths can help a lot. Whenever things get too much, stop for a bit. Center yourself and breathe in deeply.
3. **Challenge Negative Thoughts:** Sometimes our minds play tricks. Negative thoughts make things worse. Ask yourself—are these thoughts real or just based on worries? Swap out those pesky ideas for balanced and realistic ones.
4. **Use Positive Self-Talk:** Talk to yourself in a nice way to boost how you feel about yourself. Say affirmations like, "I can manage this" or "I am strong and capable."
5. **Engage in Relaxation Activities:** Find ways to chill out and feel good again—take a walk, listen to your favorite tunes, or dive into a hobby you love. Spending time on yourself can help you feel better and less stressed overall.

Tips for Helping Yourself and Others in Tough Times

When things get tough, it's important to support yourself and others. It helps keep everyone feeling OK and strong. Here are some friendly tips:

1. **Listen Carefully:** When someone is upset, just listen closely. Show you care. Don't judge. Let them know they're not alone. Say something like, "I understand how you feel."

2. **Give Practical Help:** Look for little ways to help out. Maybe offer food, a place to stay, or a ride somewhere. Small acts of kindness mean a lot.
3. **Solve Problems Together:** Help people find ways to fix their problems. Encourage them to take steps on their own too. Give advice, but also trust that they can do it.
4. **Be There Emotionally:** Offer comforting words to those who are struggling emotionally. Tell them it's OK to feel down and remind them they can get through it.
5. **Seek Professional Help When Needed:** If someone is deeply troubled or facing big mental health issues, suggest that it's a good idea to see a therapist or counselor. Offer help finding resources and support them in taking that step.

24. Survival Mindset & Mental Resilience

Survival psychology dives deep into what really happens in our minds when we're up against tough times. It's all about staying firm and pushing through challenges, no matter what. Grasping the psychology of survival means knowing how our brains handle stress, fear, and all the unknowns.

- **Fight, Flight, Freeze, or Faint:** When trouble hits, our bodies react with one of these basic responses. It's how our brain gears us up to either face the threat head-on, run away from it, freeze right there, or sometimes even faint to dodge more harm.
- **Emotional and Physical Response:** Fear sets off a bunch of changes in our bodies—like faster heartbeats and quick breaths. We become super alert, and muscles get tense. Knowing these signs is key for preppers to control their reactions in survival moments.

THE SIGNIFICANCE OF MENTAL TOUGHNESS FOR PREPPERS

In prepping, both physical skills and gear are really important, but without resilience, they can only help so much. Mental resilience is like the building block for everything else. It helps you adapt, improvise, and deal with challenges.

- **The Power of Positive Thinking:** Staying positive in tough times can make a big difference. It's not just a cliché! Studies show that folks with a strong mindset handle stress better and tackle problems more effectively.
- **Decision-Making Under Pressure:** When it comes to survival, sometimes you gotta make quick decisions. It can be the difference between life or something worse. Mental resilience helps preppers stay cool, think straight, and make smart choices—even when things get intense.

THE SCIENCE OF HAPPINESS IN SURVIVAL SITUATIONS

Natural Happiness vs. Synthetic Happiness

Happiness, when it comes to survival, shows up in two different forms: natural happiness and synthetic happiness.

- **Natural happiness** is the joy we feel when our wishes are met—like when we get what we want.
- **Synthetic happiness** kicks in when things don't go our way. Our minds still find a way to make us feel satisfied and content.

Evolutionary Survival Mechanism: Synthetic happiness is part of our brain's survival tricks. When tough times knock on our door, our brains use synthetic happiness to guard our mental health and boost our strength.

Adaptability and Flexibility: Using synthetic happiness helps you roll with the punches and flourish in hard times. Instead of lamenting what's missing, you zero in on what's right there and find joy in the now—an essential mindset for making it through any ordeal.

Harnessing Synthetic Happiness for Mental Resilience

- **Acceptance and Adaptation:** Embracing your situation and finding joy even when things go south is the secret sauce of synthetic happiness. Don't push against or dislike your circumstances. Accept them and focus on how you can adjust and tackle the hurdles.
- **Gratitude Practice:** Counting your blessings helps big time with creating artificial happiness. Even when things are rough, pause each day to reflect on your good fortunes. This small habit can toughen you up and shift your view.

COPING WITH FEAR AND ANXIETY

Fight, Flight, Freeze, or Faint: Understanding Fear Response

- **Understanding the Fear Response:** When danger or stress crosses our path, our bodies spring into action, often called the "fight or flight" response. This primal instinct is wired deep in our brains and bodies, preparing us to either tackle the threat head-on or race away from it as fast as we can. But, there are more ways we respond besides just fighting or fleeing. Sometimes, when the danger feels overwhelming, our bodies might freeze up—just stuck in place, unable to act. In extreme cases, some folks might even faint.
- **The Purpose of the Fear Response:** The fear response is crucial for survival, helping us stay safe in dangerous times. When triggered, a rush of adrenaline and other stress hormones floods through us, making our bodies more alert and ready to react quickly to possible threats. This alert state provides us with the energy and focus needed to handle tough situations effectively.

Strategies for Controlling Fear and Anxiety in Survival Scenarios

- **Regulate Your Autonomic Nervous System:** First, let's talk about keeping your body's autopilot in check—your autonomic nervous system! It handles things like heart rate, breathing, and digestion without you even thinking about it. If you learn to control this system, you'll manage stress way better.
- **Deep Breathing:** To soothe anxiety and calm your nervous system, try deep abdominal breathing. Find a comfortable spot, whether sitting or standing. Take slow, deep breaths, filling your belly, not just your chest. Inhale deeply, hold for four seconds, exhale gently for four seconds, and hold again for another four. Repeat this a few times to feel peace and relaxation wash over you.

- **Grounding Techniques:** Feeling swamped by fear or anxiety? Ground yourself by tuning into your physical sensations and surroundings. Make sure your feet are flat on the ground. If you can, stomp your feet firmly to connect with the Earth. This helps anchor you in the present moment and can ease feelings of panic.
- **Acceptance and Ownership of Circumstances:** It's crucial to embrace your current situation fully instead of fighting it. Own it! Acknowledge that while the circumstances might be tough, they are yours to face and overcome. You got this!
- **Proactive Problem-Solving and Adaptation Strategies:** Approach challenges head-on rather than letting fear freeze you. Identify exactly what's troubling you and brainstorm some possible fixes. Focus on what's controllable and act swiftly to tackle any threats or obstacles.
- **Synthetic Happiness:** Create your own happiness! Even when life isn't perfect or rewards aren't flowing in, you can still feel content and fulfilled. Your mind has the power to generate its own sense of well-being regardless of external circumstances.
- **Innovative Adaptation:** Think creatively when facing tough situations. See challenges as learning opportunities rather than insurmountable barriers. Face obstacles with readiness to adapt, resilience, and a willingness to pivot if needed.

DEVELOPING MENTAL STRENGTH THROUGH CHALLENGES

Uncovering Hidden Lessons in Adversity

- **Spot the Lessons:** When facing tough times, having a mindset of growth and learning is super important. Think of challenges as chances to grow instead of setbacks. Take a moment to consider what's happening and see if there are any lessons to be learned. You might reflect on how you reacted, figure out what went wrong, and ponder how to handle things differently next time.
- **Gaining Perspective:** Hard times often give us a unique viewpoint that can be really eye-opening. Give yourself some time to step back and see the bigger picture. Ask yourself, "What can I learn from this?" and "How does this fit into the broader context of my life?" This way, you'll find meaning in the troubles you're facing and understand their importance better.
- **Welcoming Growth Opportunities:** Adversity isn't just about making it through; it's about thriving. Embrace the chance for personal growth that comes with tough times. Use these experiences as a push for positive change, whether it's learning new skills, building resilience, or discovering more about yourself. Face each challenge with an open mind and a desire to learn, and you'll become stronger and more resilient in the end.

Embracing & Adapting to Tough Situations

- **Accept & Own It:** When tough times hit, it's super important to face reality as it is. Take charge of your life. Don't get stuck thinking about "what if" or wishing things were different. Focus on what you can control and tackle the situation head-on. By owning your circumstances, you give yourself the power to make positive changes and move forward with confidence.
- **Adjusting to Change:** Being able to roll with the punches is key when dealing with hard times. See change as a chance to learn and grow, not something to fight against. Stay flexible and be ready to change your approach if needed because what worked before might not work now. Meet new challenges with curiosity and openness. Remember, every experience teaches us something valuable and helps us grow.
- **Building Resilience:** To handle tough situations with strength, you need to build resilience. Create a support network, learn ways to cope, and take care of yourself. Try to stay positive, find meaning in hardships, and keep your sense of direction. Prioritizing mental and emotional health will help you deal with life's challenges and come out even stronger!

DEVELOPING A SURVIVAL MENTALITY

Acceptance & Ownership of Circumstances

- **Accept Your Situation:** When life tosses you a curveball, the first thing is to accept what's happening. Yeah, it can be tough. Look at your situation for what it is. Complaining or dwelling on the negatives won't fix anything. Embrace where you're at. Remember, you actually have the power to find a way through it.
- **Assume Full Ownership:** Own it completely. Your situation is yours alone—nobody else can fix it for you. By taking ownership, you're giving yourself the power to take charge and make changes. It's not always about liking where you are, but stepping up and taking control of it.

Proactive Problem-Solving & Adaptation Strategies

- **Work with Your Situation:** This means changing how you view problems and challenges. Don't just react—be proactive in looking for solutions. Seek opportunities within your current situation and adapt as needed.
- **Get Creative and Proactive:** Challenges? Get imaginative! Think outside the box and try different avenues. Jump in, look for growth opportunities, take initiatives, and face challenges head-on.
- **Exercise Your Adaptation Muscle:** Being adaptable is super important. To build resilience, practice adjusting to all kinds of situations and changes. Embrace these as chances to learn and grow. Your adaptability will strengthen each time you practice.

25. Survival Strategies for Natural Disasters

In nature, survival is frequently dependent on preparation and strategic action. This chapter provides recommendations and a thorough path for overcoming the most formidable obstacles faced by hurricanes, earthquakes, floods, wildfires, and other natural disasters.

SURVIVING A HURRICANE

Let's take a look at what to do when a hurricane's on the way:

Finding Secure Shelter: Your top priority is finding a safe spot. Look for buildings built to handle strong winds. The best protection comes from sturdy buildings with strong walls and foundations. Try to find interior rooms without windows, like bathrooms, closets, or hallways. They'll keep you safe from flying debris and flooding. Remember, low areas can flood easily, so move to higher ground if you need to leave.

Seeking Solid Protection: Stick to lower levels or reinforced areas that can handle the storm's power. Avoid rooms with big windows or glass doors—they might break. Stay away from outside walls and flood zones to keep safe. Use heavy furniture or sturdy fixtures as extra shields against flying objects and any potential collapses.

Staying Informed: When things get wild, information is your best friend. Have a battery-powered weather radio ready for updates. Keep an ear out for weather forecasts and alerts from local authorities. Watch weather patterns and any evacuation orders closely so you can act quickly if needed.

Fortifying Your Home: Get those storm shutters up or board up windows tight to protect against broken glass and flying debris. Reinforce your garage doors and entryways to keep them from failing. Clear out loose items outside that might become projectiles in the wind. Do everything you can beforehand to protect your home before the storm arrives.

Taking Immediate Action: If local authorities tell you to evacuate, do it right away! Don't hesitate—go to designated shelters or sturdy buildings fast. Be ready to follow emergency plans and routes as things change quickly.

Leaving Mobile Homes: If you're in a mobile home, leaving is super important when a hurricane's coming. Mobile homes are very vulnerable in these storms. Evacuate early and find safer shelter ahead of time.

SURVIVING A TORNADO: ENDURING THE TWISTER'S FURY

Here's what you need to know to increase survival during a tornado:

Seek Sturdy Shelter: Just like with hurricanes, finding a strong shelter is key, but you've got way less time to get there. Look for tough buildings that can handle the tornado's wild winds. Basements, storm shelters, or interior rooms with no windows are your best bet. They offer significant protection against the storm's fury. Aim for places built with reinforced concrete or steel because they stand firm against a tornado's punch.

Position Yourself Wisely: Never try to hide in vehicles, trailers, or weak structures that could be easily lifted by the tornado's winds. Choose low-lying spots or dips in the ground, as they naturally shield you from the storm's rage. Lie flat and cover your head with your arms to avoid flying debris and increase your chances of survival.

Follow Protocols: Keep track of alerts from local authorities and weather forecasts. When tornado warnings are issued, act quickly. Find safe zones without wasting any time. Make sure you're familiar with community evacuation plans and emergency procedures so you know where to go.

Don't Drive Unless Needed: Unless it's absolutely necessary, avoid driving during a tornado threat. Tornadoes can appear suddenly and put drivers in danger before they realize it. If you're caught on the road when a tornado hits, leave your car and find strong shelter or a low-lying area immediately. Don't try to outrun the tornado; that usually makes things worse.

Put Safety First: Stay close to the ground. Head for the lowest level of a sturdy building or a natural depression. Cover yourself with thick blankets, mattresses, or cushions to protect against flying debris and injuries. Keep a battery-powered weather radio and emergency supplies handy so you're always informed and ready for anything the storm throws at you.

TSUNAMI PREPAREDNESS: EMBRACING RESILIENCE IN THE FACE OF NATURE'S WRATH

Tsunamis can be just as suddenly terrifying as a tornado. When the ocean swells and those huge waves rear up, it might already be too late to escape. This guide dives into some tips and tricks that can help you out if you ever face this situation.

Immediate Action

When the ground shakes under your feet and the ocean pulls back with a scary noise, you have to act right away. Don't waste time—get to higher ground quickly. Earthquakes often come before a tsunami, so move fast when you feel those tremors. Your safety is super important, so don't think twice about evacuating.

Seeking Elevated Structures

High ground? That's your best bet for safety. Look for tall structures or climb to the highest points you see around. Get up as high as possible, where the tsunami waters are less likely to reach and where you have a better shot at staying safe.

Planning Evacuation Routes

Plan your evacuation routes ahead of time and know exactly how to get to safety before any danger shows up. Make sure you're familiar with paths leading to higher ground and have more than one way out just in case. Learn about local evacuation plans and know where the safe spots are around your area.

SURVIVING FLOODS: NAVIGATING THE WAVES OF CHALLENGE

How to Stay Safe During a Flood

In the wake of heavy rain and rising waters, floods can be a significant problem. Here's how to stay safe during a flood:

Heed Evacuation Orders

It's crucial to follow evacuation orders immediately. Why? Because escape routes can disappear quickly. Officials issue these orders to keep everyone safe, so prioritize your safety and your family's safety by evacuating as soon as possible. Waiting can be dangerous since floodwaters rise rapidly and can trap those who delay.

Avoid Swimming

Don't attempt to swim in floodwaters! The water conceals many dangers, such as debris and strong currents. Instead, find safer ways to navigate through flooded areas to avoid these hidden risks.

Swimming Prudently

Sometimes, you might have no choice but to swim. If that happens, be extremely cautious! First, assess how deep and fast the water is before entering. Make sure it's as safe as possible. Swim with the current, not against it, to conserve energy and maintain control. Strong currents can overpower even the best swimmers.

Seek Higher Ground

As soon as floodwaters begin to rise, move to higher ground immediately. Seek out elevated areas and stay away from the water. Look for sturdy buildings or natural high spots that can provide shelter.

Secure Essential Supplies

When flooding occurs, accessing basic supplies can become difficult. Ensure you have enough bottled water, non-perishable food, essential medications, and a flashlight. It's also wise to have a battery-powered radio and a first-aid kit readily available.

SURVIVING AVALANCHES: MASTERING THE SNOWY ONSLAUGHT

Avalanches: Quick Thinking & Action

Avalanches demand quick thinking and fast action. Here's what you need to know to boost your chances of making it out alive:

Moving to Side Slopes

First thing, move to the sides of the slope, where snow gathers less. By sticking to the edges, you reduce the chance of getting buried by the avalanche. Act quickly and confidently, always prioritizing safety.

Utilizing Anchors

Find any anchors around you. Grab onto sturdy trees, rocks, or anything that sticks out to stop yourself from being pulled away by the snow. Hold on tight until it's safe to let go without getting dragged under.

Creating Air Pockets

If you get stuck in an avalanche, create air pockets as soon as the snow stops moving. Use your hands, arms, or whatever you have to dig out space around your head and face. This will help ensure you have air to breathe.

WILDFIRE SAFETY: SURVIVING NATURE'S BLAZE

As wildfires become increasingly common, it is critical to know how to prepare for them. They behave similarly to hurricanes in that they usually provide some notice. But wildfires are also extremely hazardous and destructive. Here's how to prepare for a wildfire.

Ensuring Breathing

When faced with the smell of wildfire smoke, cover your nose and mouth with a wet cloth or mask. This helps filter out harmful particles in the air. If possible, stay inside a sealed building to breathe more easily. Keep your lungs safe so you can think clearly and move around.

Seeking Shelter

Seek out buildings made from fire-resistant materials. Avoid areas with lots of vegetation or flammable objects. Stay in open spaces away from potential fire sources to reduce the risk of being caught in the flames.

Using Vehicles Wisely

If authorities advise evacuation, leave quickly using your car. Be mindful of avoiding crowded roads or potential blockages. If evacuation is not possible, use your vehicle as a last resort shelter against the flames. Stay alert and be prepared to adjust your plans if necessary.

Creating Firebreaks

Create firebreaks to slow the fire's progress. Clear vegetation and debris around your home using simple tools like shovels and rakes. These barriers can help delay the fire, giving you more time for help to arrive or to evacuate. Work together with neighbors and local authorities to create larger firebreaks, combining efforts to fight off the flames.

GENERAL SURVIVAL STRATEGIES

Knowing what's happening is super important when things are uncertain. Keep up with the news using a weather radio runs on batteries & reliable sources. Watch weather forecasts and official alerts for any threats coming your way. Timely lets you make smart choices and take steps to protect yourself and your loved ones. After a disaster, you need to be ready to take care of yourself for at least 72 hours without help outside. Gather the basics and get ready to face tough conditions. You'll need to use your smarts & toughness to get through it.

26. Herbal Medicine Essentials

Introduction to Herbal Medicine for Preppers

Herbal medicine has been around for thousands of years, using natural remedies from plants to tackle various health issues. Sometimes, when you can't get to a doctor, knowing how to use plants for their healing powers can be really important.

At its core, herbal medicine treats not just the symptoms but also the root causes of health problems. Unlike man-made drugs, herbal therapies often work with your body to boost overall health and vitality. Plus, using plants can be more eco-friendly and sustainable.

Learning about herbal medicine allows you to identify and grow plants for their health benefits.

Understanding Herbal Remedies and Their Benefits

Herbal remedies come in different forms. There are teas, tinctures (which are like extracts), poultices (which are pastes), and salves (like ointments). Each type is made to extract and concentrate the beneficial properties of plants to help with specific health issues.

One big plus of herbal remedies is their flexibility. Whether you're dealing with a tummy ache, a cough, or need a boost to your immune system, there's usually an herb that can help and be effective. They also tend to have fewer side effects compared to synthetic medications, making them a gentler choice for many people.

Moreover, many herbs offer additional benefits, such as being antioxidants or having anti-inflammatory and antimicrobial properties. Incorporating these herbs into your daily routine not only addresses immediate health issues but also strengthens your resilience against illness over time.

Building Your Herbal Medicine Kit

Creating a thorough herbal medicine kit is essential if you want to incorporate herbal remedies into your prep plan. Here's what you need:

Research First: Start by digging into the medicinal qualities of different herbs and how they work for specific health issues. Consider factors like climate, where they grow best, and how easy they are to cultivate.

Stock Up: Gather a variety of herbs, both dried and fresh, to cover all your health needs. It's wise to stock up on common herbs with many uses, as well as some specialty ones for specific needs.

Stay Organized: Your herbal kit should be tidy and easy to access. Make sure herbs are clearly labeled and stored in a cool, dark spot to maintain their potency. Consider buying containers, jars, and labels to stay organized and efficient.

METHODS FOR PREPARING HERBAL REMEDIES: INFUSIONS, DECOCTIONS, TINCTURES, AND BEYOND

Here is a closer look at the methods utilized in herbal therapy:

- **Infusions** involve soaking herbs in hot water, which helps release their beneficial compounds. To make an infusion, place fresh or dried herbs in a heatproof container, pour boiling water over them, cover, and let it steep for 10 to 15 minutes. Then, filter and drink. Infusions work well with soothing herbs like chamomile and peppermint.
- **Decoctions** are similar to infusions but stronger. You boil plants in water to extract more potent chemicals. For a decoction, put herbs in a saucepan with cold water, bring it to a boil, then lower the heat and simmer for 20 to 30 minutes. Strain it before drinking. This method is ideal for tougher plant parts like seeds, bark, and roots.
- **Tinctures** are made by soaking herbs in alcohol or vinegar to create concentrated extracts. To make a tincture, chop up some herbs and place them in a glass jar. Pour alcohol or vinegar over the herbs, seal the jar, and shake it daily for a few weeks to keep everything mixed. After straining out the solids, store the liquid in dropper vials. Tinctures are handy for precise dosing and long shelf life.
- **Other Methods**: There are many other ways to use herbs, including poultices, salves, syrups, and capsules. Experiment with different methods to find what works best for you.

ESSENTIAL HERBS FOR SURVIVALISTS: IDENTIFICATION, BENEFITS, AND APPLICATIONS

You know, you might just find yourself gravitating toward a few essential herbs. Each one has its own special traits and uses. Their medicinal qualities are so useful too! These herbs find places in our kitchens and in our home remedies. You'll see they're incredibly versatile.

Basil: The Everyday Marvel Herb

Basil has such inviting leaves with amazing flavors. And oh, the benefits! It's great for minor cuts because of its antibacterial powers. Plus, it helps with belly aches too. Whether you sip it as tea or add it to your food, basil brings both taste and health perks to the table.

Coriander/Cilantro: The Tangy Healer

Also called cilantro, this herb is packed with antioxidants. It helps detox the body and boosts overall health. Its antimicrobial traits make it a warrior against foodborne bad guys. You can sprinkle it on your dishes or make cozy infusions; coriander shines both as a medicine and a tasty ingredient.

Kaffir Lime: The Citrus Gem

The lovely leaves of the kaffir lime tree are super fragrant. They're wonderful in herbal remedies! With antimicrobial and anti-inflammatory characteristics, these leaves can help with digestion troubles and fevers too. Brew them in teas or infusions for a refreshing drink that keeps you feeling fine.

Thyme: Aid for Respiratory and Digestive Health

Thyme, with its tiny leaves and lovely smell, is awesome for helping your breathing and tummy feel better. Thanks to thymol (a special part of thyme), it eases sore throats & tummy aches. You can make thyme tea or add it to your favorite dishes.

Lemon Verbena: Calming Herb for Stress Management

Lemon verbena, which smells like lemons and feels so calming, is perfect for stress relief. The citrus part helps you relax and keeps your digestion happy. Drink it as tea or mix it into herbal blends—lemon verbena will surely help you unwind & keep your gut smiling.

Dill: Digestive Aid and Flavor Enhancer

Dill has feathery leaves and a taste you'll love. It's not just tasty but also great for digestion. Carvone & limonene (fancy names, huh?) work wonders for your tummy. Sprinkle dill on salads or use it as seasoning—it's delicious and helpful!

Parsley: Nutrient-Dense Herb for Overall Health

Parsley, bursting with green goodness and packed with nutrients, is fantastic for staying healthy. Full of vitamins, minerals, & antioxidants—it helps your kidneys and cleans out bad stuff from your body. Use parsley as a garnish or make tea—you'll get essential nutrients easily.

Bay Leaf: Fragrant Addition to Herbal Remedies

Bay leaf has a lovely smell and gentle taste. With cineole inside (a magical part of bay leaf), it helps you breathe better & aids digestion. Pop bay leaves into soups, stews, or herbal drinks—they're superb for your respiratory health and digestion!

Mint: Refreshing Herb for Digestive and Respiratory Wellness

Mint, with its refreshing taste and cool vibe, works wonders for your belly and breathing. Why? It contains menthol, which helps calm tummy troubles and clears up stuffy noses.

Chives: Immune-Strengthening Herb for a Prepper's Nutrition

Chives have a mild onion taste that's really nice. Plus, they're great for your system. Full of vitamins and minerals, chives help keep you healthy. Toss them into salads, soups, or even make a tea. Your body will thank you!

Curry Tree: Key Component in Herbal Remedy Formulations

The curry tree has leaves that smell amazing and taste unique. It's super useful in herbal medicine, thanks to its germ-fighting and anti-swelling properties. Use these leaves in your cooking or make a tea that's both delicious and good for you.

Rosemary: Cognitive-Enhancing and Antioxidant-Rich Herb

Rosemary smells like pine trees and packs a strong flavor punch. It's got great benefits too! This herb can help improve memory and is full of antioxidants. Make rosemary oil, sip it in tea, or add it to your meals. It's a great way to boost brain health!

Chili: Spice for Pain Relief and Improved Circulation

Chili peppers, known for their hot kick and powerful medicinal traits, might seem tough, but they surprisingly provide pain relief and circulation perks. The capsaicin in them eases pain and helps blood flow, making chili peppers super valuable in herbal remedies.

Mustard: Versatile Herb for Both Medicinal and Culinary Applications

Mustard, with its tangy taste and many uses, shines as a multi-purpose herb perfect for both medicine and cooking. The glucosinolates in mustard offer antimicrobial and anti-inflammatory benefits, which support better health. You can use mustard seeds when cooking or make poultices for the skin.

Fennel: Herb for Digestive Comfort and Respiratory Health

Fennel is loved for its sweet seeds and light licorice flavor. It's great for digestion and helps with breathing issues. The anethole in fennel calms tummy troubles and clears up stuffy noses. You can brew some fennel tea or add it to your recipes.

Garlic: Powerful Antimicrobial and Immunity-Enhancing Herb

Garlic stands out with its strong smell and mighty health benefits. It's a top-notch antimicrobial and a great immune booster. Thanks to allicin, garlic fights germs and helps keep your immune system strong. You can eat garlic raw or mix it into your meals for a health boost.

NATURAL REMEDIES FOR EVERYDAY ILLNESSES AND EMERGENCIES

Herbal medicine provides a natural and handy way to deal with everyday ailments and sudden issues. Because of this, it is a great tool for taking care of yourself and being ready for anything. By learning about different herbs and their benefits, you can create special remedies for many health problems.

Here's how you can use herbal medicine for some specific issues:

- **First Aid:**
  Herbs can be super important in first aid, helping with small cuts, bruises, and pains.
    - *Comfrey and calendula* help heal wounds and reduce swelling.
    - *Arnica* is famous for easing pain and bruising, making it a must-have in natural first aid kits.

- **Cold & Flu:**
  During cold and flu season, herbal medicine can support your immune system and alleviate symptoms.
    - *Echinacea* is a go-to for boosting immunity, helping to shorten the duration of colds or respiratory infections.
    - *Elderberry syrup* is loved for its ability to fight viruses and reduce the severity of flu symptoms.

- **Digestive Issues:**
  Many herbs are effective in treating tummy troubles like indigestion, bloating, and gas.
    - *Peppermint* soothes an upset stomach and eases symptoms of irritable bowel syndrome (IBS).
    - *Ginger* improves digestion, eases nausea, and reduces inflammation in the gut.

- **Pain Relief:**
  For pain relief, herbs offer natural alternatives to conventional pain medications.
    - *Willow bark* works like aspirin due to its salicin content, making it good for headaches, muscle pain, and arthritis.
    - *Turmeric* has strong anti-inflammatory properties that help with chronic pain from conditions like rheumatoid arthritis and osteoarthritis.

- **Respiratory Issues:**
  Herbs can help maintain lung health and ease conditions like asthma, colds, and coughs.
    - *Thyme* helps clear congestion and fight respiratory infections with its expectorant and antibacterial qualities.
    - *Mullein leaf* soothes the respiratory tract, calming coughs and clearing bronchial congestion.

Anxiety & Stress:
- Herbal remedies can be gentle allies when dealing with anxiety and stress, helping you relax and feel better emotionally.
    - *Lavender*, for instance, is well-loved for its calming smell. Often found in teas, tinctures, and essential oils, it can help you unwind and even sleep better.
    - *Lemon balm* is another good option. It can help you feel less stressed and more at peace.

Skin Conditions:
- Herbs to the rescue! They have healing properties that work well for all sorts of skin issues—eczema, psoriasis, acne, and even minor wounds.
    - *Calendula* is pretty amazing with its anti-inflammatory and antimicrobial powers, making it perfect for minor cuts, burns, and rashes.
    - *Chamomile* is known for its gentle touch on the skin. It's often used in creams, ointments, and baths to calm irritations and speed up healing.

Insomnia & Sleep Disorders:
- When it comes to sleep problems, herbal medicine offers natural alternatives to sleeping pills. These herbs help you sleep well without the risk of dependency or nasty side effects.
    - *Valerian root* is a favorite for treating insomnia and enhancing sleep quality due to its sedative qualities.
    - *Passionflower* is great too! It helps calm your nervous system, reducing anxiety so you can relax before bed.

Allergies & Hay Fever:
- Herbal remedies can really help with seasonal allergies and hay fever symptoms. They reduce congestion, sneezing, and itching, making life more comfortable.
- **Nettle leaf** works wonders—it's a natural antihistamine that helps with inflammation and blocks histamine release.
- **Eyebright** is amazing too! It soothes irritated eyes and nasal passages, truly a lifesaver for those with allergies.

Urinary Tract Infections (UTIs):
- Some herbs have special antibacterial and diuretic powers that can both prevent and treat urinary tract infections.
- **Cranberry** is an old favorite. It stops bacteria from clinging to the urinary tract lining and keeps the urinary tract healthy.
- **Dandelion leaf** is super helpful too. It's a gentle diuretic that helps flush out bacteria and ease UTI symptoms.

Menstrual Cramps and Menopausal Symptoms:
- Herbal medicine offers natural options for handling menstrual cramps and alleviating menopausal symptoms like hot flashes and mood swings.
- **Dong quai** is popular in traditional Chinese medicine. It helps regulate menstrual cycles, relieve pain, and ease menopausal woes.
- **Black cohosh** is another great herb! It's known for balancing hormones, which helps reduce hot flashes and other discomforts.

Stomach Upset & Food Poisoning:
- Herbs can provide significant relief for stomach upset, nausea, and food poisoning. They soothe the digestive tract and reduce inflammation.
- **Peppermint tea** is an old but gold remedy! It eases stomach cramps, bloating, and nausea, thanks to its calming properties.

Activated charcoal is handy too. It absorbs toxins and bad stuff in your gut, helping to tackle food poisoning and diarrhea.

27. Tools for Off-Grid Survival

OVERVIEW OF ESSENTIAL PREPPING EQUIPMENT AND TOOLS

Understanding the Role of Tools in Prepping

Tools are the backbone of any prepper's kit. They are essential instruments for survival in tough times. From cutting tools to lighting gear, each item plays a crucial role in making sure you're prepared for anything.

Importance of Versatility & Reliability

A tool that does many jobs well and can take a beating is priceless. It has to handle tough situations.

Cutting Tools

- **Knives – Types, Features, & Uses**

    Knives come in many types, such as fixed blade and folding. Each kind has special features. You can use knives for cutting, slicing, or carving. They're necessary for food prep, building shelters, and self-defense.

- **Axes and Hatchets – Versatile Tools for Survival**

Axes and hatchets are very handy tools. You can chop wood, hammer stuff, and split logs with them. They're small but mighty! Perfect for surviving in the wilderness and building shelters.

- **Machetes – Practical Utility in Various Situations**
  Machetes are super useful too! They cut through vegetation, thick brush, and help make tools on the fly. You need them for clearing paths, gathering firewood, and yes, even self-defense in dense woods.

Lighting Equipment

- **Flashlights – Brightness, Battery Life, & Durability:** Flashlights give you light when it's dark, making them essential for nighttime activities and emergencies. Look for models with high lumens, long battery life, and a sturdy build to ensure reliability.
- **Lanterns – Illumination for Campsites & Shelters:** Lanterns are perfect for lighting up campsites and shelters, creating a bright and cozy atmosphere. Choose models with adjustable brightness and long-lasting LED bulbs for optimal use.
- **Headlamps – Hands-Free Lighting Solutions:** Headlamps are incredibly handy because they free up your hands, making them ideal for tasks that require both hands. Opt for headlamps with adjustable straps, multiple light modes, and water-resistant designs to maximize their usefulness outdoors.

Fire Starting Devices

- **Lighters – Convenience & Portability:** Lighters are easy to carry and excellent for starting fires with tinder and kindling. Keep some waterproof and windproof lighters in your kit to ensure you can start a fire even in tough conditions.
- **Ferro Rods – Reliable Sparking Tools:** Ferro rods are great because they produce hot sparks when struck against a rough surface, igniting materials quickly. Include a ferro rod in your kit as it doesn't require fuel to start a fire.
- **Fire Starters – Tinder & Kindling Options:** Fire starters make it easy to get your fire going. Items like cotton balls coated with petroleum jelly or special fire sticks work well. Pack these lightweight and waterproof fire starters for emergencies.

Shelter Building Supplies

- **Tarps & Tents – Handy Shelter Solutions:** Tarps and tents are great for creating portable shelters to protect you from the weather. Pick ones that are tough and waterproof with strong seams so they last longer.
- **Cordage – Useful Ropes for Building and Securing:** Cordage like paracord or nylon rope is really handy for building shelters, tying up gear, and doing all sorts of tasks. Make sure to carry different types and lengths of cordage so you're ready for anything.
- **Emergency Blankets – Keeping Warm in Tough Conditions:** Emergency blankets, sometimes called space blankets, reflect your body heat to keep you warm in cold weather. They're lightweight and easy to pack, so throw a few in your kit for extra warmth.

Navigation Instruments

- **Compasses – Helping You Find Your Way:** Compasses are super helpful when you're out in the wilderness. They help you figure out where you are and which way to go. Learn some basic compass skills and make sure to pack a reliable one in your gear.
- **Maps – Must-Haves for Finding Your Path:** Maps are crucial for planning your route and finding your way, whether you're in the city or out in nature. Get detailed maps of your area and get familiar with local landmarks & features.
- **GPS Devices – High-Tech Navigation Helpers:** GPS devices give you precise location info and help guide your route, making navigation easier even in remote spots. Keep a GPS device with good maps on hand, along with extra batteries just in case.

Water Purification Systems

- **Water Filters – Portable Filtration Devices:** Water filters take out the nasty stuff and germs from water, making it safe to drink. It's good to get a portable water filter with a large capacity and changeable filters for long-term use.
- **Water Purification Tablets – Chemical Treatment Methods:** Water purification tablets, like chlorine dioxide or iodine, can eliminate bacteria and viruses in water. It's smart to keep some tablets as a backup method for purifying water in case of emergencies.
- **Boiling Equipment – Traditional Sterilization Techniques:** Boiling water is highly effective for killing harmful microorganisms. Bring a lightweight and durable pot or kettle to boil water over an open flame or camp stove. It's a simple and reliable method!

First Aid and Medical Supplies

- **First Aid Kits – Basic Medical Supplies for Emergencies:** First aid kits contain essential items for treating cuts, bruises, and minor illnesses during emergencies. Ensure your kit includes bandages, wipes, gauze pads, and other basic medical supplies.

- **Trauma Kits – Advanced Wound Treatment Tools:** Trauma kits are invaluable for more serious injuries, equipped with special supplies like tourniquets, hemostatic dressings, and chest seals. Include these in your kit for comprehensive wound care.
- **Medications – Essential Drugs for Health Maintenance:** Medications like painkillers, allergy pills, and antibiotics are crucial for maintaining health and managing symptoms. Make sure to pack a sufficient supply of essential medications in your kit to be prepared for emergencies.

Food Preparation Equipment

- **Camp Stoves:** Camp stoves are a super handy way to whip up meals outdoors. Just connect the fuel source, light up the flame, and place your cookware on the stove. With adjustable heat settings, you can simmer, boil, and fry dishes without any hassle. It's that easy.
- **Cooking Utensils:** Can't cook without the right tools! From spatulas to pots and pans, cooking utensils are essential for making and serving food outdoors. These tools help create delicious meals over an open flame. Go for lightweight and durable utensils to make packing and cleaning a breeze.
- **Food Preservation Supplies:** Keep your provisions fresh on long trips with food preservation supplies. Vacuum-sealed bags (amazing), dehydrators, and canning jars all help keep food longer. Preserve fruits, vegetables, and meats for a well-stocked pantry (in any situation).

Communication Devices

- **Two-Way Radios: Handy Gadgets for Group Coordination**

    Two-way radios help you chat in places where cell phones don't work. Pick ones with long-range signals, many channels, and weather alerts. Stay linked up, coordinate moves easily, and share vital info without a hitch.

- **Emergency Radios: Weather Updates and Alerts That Matter**

    Emergency radios give you weather updates when it counts. Tune to NOAA Weather Radio frequencies for real-time forecasts and evacuations. Stay alert, stay in the know, and stay safe with an emergency radio by your side.

- **Signal Whistles & Mirrors: Emergency Signaling Devices**

    Signal whistles and mirrors get attention fast in emergencies. A loud whistle or a mirror's flash tells rescuers where you are. Keep these handy items close for peace of mind when things go south.

Personal Protection Gear

- **Protective Clothing: Gear for All Weather**

    Protective clothing keeps you safe from the weather—warm and dry. Choose fabrics that wick away moisture to control temperature and avoid hypothermia or heat illness. Stay cozy and protected on your journey.

- **Masks and Respirators: Shield Against Airborne Dangers**

    Masks and respirators guard against bad air like pollutants and germs. Pick masks that fit snugly over your nose and mouth to filter out harmful particles. Breathe easy and keep health risks at bay in tough spots.

- **Self-Defense Tools: Gear Up for Security**

    Self-defense tools keep you safe when things get sketchy. From pepper spray to tactical knives, pick tools that match your skills and comply with local laws

PROJECT 28: BUILD A DIY BICYCLE GENERATOR

It may sound very old school, but you can actually generate your own power. It isn't the most effective or efficient method, but it is easily the healthiest. A bicycle generator will give you a good work out along with power.

Materials and Preparation

Before you start, you will need the following materials:

- Bicycle
- Diode
- V-belt
- Inverter
- Motor
- Batter 2 x 4 in wood
- Wood screws
- Hammer/screwdriver
- Tap measure
- Perforated plumbers steel

Steps

1. Create a stand to raise the bike. The design is up to you, with the only requirements being that it needs to be stable (the bike won't wiggle on it) and it needs to keep the bake half of the bike between 5 and 7 in off the ground.
2. Remove the back tire of the bike.
3. Put the bike on the stand and add the V-belt on the back rim.
4. Add the motor to the back end of the stand. It needs to be tightly secure, and it should spin in the same direction as the belt.

Bike Stand with the Battery Attached

5. Create a series with the diode so that the current flows from the motor on the back to the battery. This means that the cathode should point toward the positive terminal.
6. Connect the diode and battery.
7. Connect the inverter and battery leads.

The Battery and Inverter Connections

8. Test the setup. It should look something like the following picture.

PROJECT 29: GET POWER FROM DEAD BATTERIES

During an disaster, everything you have counts, and that includes dead batteries. Most of them have a bit more left, it's just a matter of milking them a little bit more.

Materials and Preparation

Before you start, you will need the following materials:

- 2 dead batteries
- Full metal barbecue clamp

Do *not* use broken, corroded, or leaking batteries.

Steps

These steps are for resetting a rechargeable batter that doesn't seem to want to charge.

1. Clean the barbecue clamp to make sure there isn't any grease or food particles on it.
2. Put the batteries end to end, the negative (-) sides touching.
3. Put between the two prongs of the barbecue clamp and squeeze them together for 30 seconds.
4. Place the rechargeable battery in the charger, and it should recharge. If not a rechargeable battery, you can see about using the battery.

Squeeze the Two Negative Ends Together.

Sometimes you will need to create series with multiple batteries to get a charge. In this case, you will need to set the positive (+) end to the negative end of the other battery. You can actually make a long series of three or four batteries to try to get more out of them. You can even create a battery pack that you can connect to a USB plug to charge a phone.

PROJECT 30: MAKE YOUR OWN WIND TURBINE FOR FREE POWER

While it will take a bit of research, you can create a wind turbine to generate some of your own energy whenever there is wind. This project may take a bit more than the others, but it can be a great source of energy during windier months.

Materials and Preparation

Before you start, you will need the following materials:

- Turbine kit
- Welding tools and protective gear

You should follow the instructions on the kit. The following steps include a bit more before you start building, as well as the general process.

Steps

1. Research when and where wind typically flows in your area. You want to make your wind turbine efficient and effective, which means having it face the directions that the wind is most likely to flow. You can do this online, or you can use an anemometer (wind measuring tool) to test the different locations on your land to see where the wind comes from the most.
2. Research applicable laws and codes regarding wind turbines in your area.
3. Determine what kind of wind turbines you want to install. The length of the blades will determine the next step.
    a. You can build your own, in which case you will need to decide the material as well since that will determine how long the blades will be.
    b. Buy the blades, in which case you will need to read all of the details and specifications so you know how long the blades are.
4. Determine how you want to space out your wind turbines on what you've found.
5. Select what kind of generator you want.

Generator for Your Wind Turbine

6. Build the spindle and pokes for your vertical axis turbine.
    a. Attach (usually through welding) the spindle to the spindle plate. If you buy a kit, this is probably already done.
    b. Slip the hub in place over the spindle.

- c. Attach the spoke flange to the hub.
- d. Connect the spokes.
- e. Insert the four studs to the upper flange.

7. Mount the magnets
   - a. Set the rotor on the studs.
   - b. Add the spacers on the studs.
   - c. Add the stator on the rotor.
   - d. Add the upper magnet rotor.

8. Complete the build.
   - a. Take the assembly off of the spindle.
   - b. Attach (usually through welding) the spindle flange to the tower.
   - c. Install the bracket.
   - d. Put the tapered roller bearing on top of the spindle.
   - e. Attach the stator and the grease cap.

9. Add the electrical components to the turbine.
   - a. Connect the battery and charge controller.
   - b. Connected the insulated wire to the controller.
   - c. Work the thread wire through the turbine's base and shaft.
   - d. Connect the battery to the turbine.

28. Building a Community

The importance of unity & working together really shows itself during a crisis. It's about more than just cooperation—it's the core of survival.

- **Introduction to Building Community:**
  Building a community isn't fancy—it's a must-have. It means growing trust, sharing values, and having a common goal. When things go wrong, these bonds are lifesavers, providing comfort, help, and togetherness when everything feels shaky.
- **Understanding Crisis Moments:**
  Every crisis is different, with its own challenges. They truly test how strong and flexible communities are. By thinking ahead and understanding potential crises, we can be better prepared and ready to tackle new threats.
- **The Power of Community Resilience:**
  Resilience is crucial in building community. It's the ability to face tough times, adapt to change, and emerge even stronger. Strong communities have tight social ties, solid support systems, and shared goals for everyone's well-being. By fostering resilience, communities can thrive during the wild ride of a crisis.

Communication Strategies During Crisis

- Being clear, timely, and accurate with communication helps bring people together. It keeps things organized and calm, allowing communities to gather resources, share information, and make plans.
- When you have reliable ways to communicate and share information openly, everyone can handle crises better. It's like having a good map - you know where you're going.

Organizing a Survivalist Community

- Putting together a survivalist community takes some planning. You need people to work together as leaders, and decisions need to be made collectively.
- Set roles clearly. Everyone should know what they need to do. Trust each other and take responsibility for your tasks. By working together and staying strong, the community will be ready for any crisis.

Things to set up:

- Who leads what
- Who does what
- What's most important

Five Key Areas of Expertise in Community Building

1. **Medical Care:** Provide health services when folks are sick or injured. Keep everyone healthy and clean.
2. **Shelter:** Set up safe and lasting places to live. Protect people from the weather.
3. **Nourishment:** Secure food and water. Cook and share it. Grow food if possible.
4. **Security:** Ensure safety from outside threats. Establish rules for everyone to follow.
5. **Communications:** Keep talking! Set up good networks to share important news and help everyone stay connected.

Minimum Effective Size of Survivalist Groups

While bigger groups may seem like a great idea because of the extra hands and resources, smaller groups often do better. They can move fast, stick together more, and work efficiently. According to how teams work best and coordinate tasks, having just five people in your survivalist group is usually enough. Each person should focus on one key skill.

Scalability of Community Building Plans

Community plans need to be flexible. Things change, threats evolve. The basic rules for keeping a community strong stay the same, but how we respond can shift based on what's happening and for how long. If we break down our strategies into parts, focus on what's most important, and adjust resources as needed, our communities can adapt well in any situation.

Equipment and Supplies for Community Preparedness

Preparedness is key. It's what keeps a community strong during tough times. By gathering essential tools and supplies, we can reduce the impact of crises and improve our ability to bounce back. Everything from medical kits to shelter materials, food rations to communication devices—they all play a super important role in keeping everyone safe and sound.

- Medical supplies and equipment
- Shelter materials and tools
- Food and water provisions
- Security equipment and weaponry
- Communication devices and batteries

Progressive Preparedness Timeline

Preparing for crises requires a systematic and incremental approach. By dividing preparedness efforts into distinct phases and milestones, communities can track their progress and prioritize their actions effectively. From short-term survival needs to long-term self-sufficiency goals, each stage of the preparedness timeline brings communities one step closer to resilience and readiness.

- **Short-term survival (3 days)**
- **Medium-term sustainability (3 weeks)**
- **Long-term self-sufficiency (3 months to 1 year)**

Team Responsibilities in a Survivalist Community

Within a survivalist community, each team plays a vital role in ensuring the well-being and security of its members. From medical care and shelter construction to food cultivation and security patrols, each team is responsible for fulfilling specific tasks and objectives outlined in the community's preparedness plan.

- **Medical team:** Providing healthcare services and first aid
- **Shelter team:** Constructing and maintaining living accommodations
- **Nourishment team:** Sourcing, preparing, and distributing food
- **Security team:** Protecting the community against external threats
- **Communications team:** Establishing and maintaining communication networks

One-Year Survival Calendar

Imagine a survival calendar that guides you step-by-step through a whole year. Each day offers something new: training exercises, equipment buys, skill learning, and even community drills. It's like building a stronger community bit by bit. Every day, you move closer to resilience and readiness.

Urgency of Preparation for Crisis Situations

In uncertain times and upheaval, preparation is key. The urgency to get ready for crisis situations is super important. Being able to respond and adapt when things go wrong can really make the difference. It's the line between survival and catastrophe. Let's focus on being prepared!

29. Off-Grid Energy Solutions

Alternative energy sources offer practical ways to power up key devices and stay safe during tough times.

A Look at Renewable vs. Non-Renewable Energy Sources

Renewable energy, coming from naturally refilled resources like sunlight, wind, and water, never runs out. These sources are everywhere, clean for the planet, and always around. That makes them great for lasting use. On the flip side, non-renewable energy like coal, oil, and natural gas is limited. We dig these up from the Earth, but once they're gone, they're gone for good.

Perks of Alternative Energy for Prepping and Surviving

- **Independence:** With alternative energy systems, you get to break free from the grid. So, when emergencies hit, you've still got power. No worries!
- **Sustainability:** These renewable sources are all about being green and lasting. They help cut down on using up limited resources and keep the Earth happy by lowering your carbon footprint.
- **Cost-effectiveness:** Yes, setting these systems up might be pricey at first. But in the long run, they'll save you money on those pesky energy bills. It's a smart move that pays off over time.

SOLAR POWER

Definition and Explanation

So, as we chatted about before, photovoltaic panels are those clever gadgets in solar power systems. They take sunlight and turn it into electrical energy. This renewable energy is a fabulous choice for folks who want to be self-sufficient. Why? It's abundant, clean, and easy to get!

Advantages and Disadvantages

Advantages:

- **Abundance:** The sun gives us endless energy! There's plenty of power for lots of different uses.
- **Cleanliness:** Using solar power means making electricity without any nasty emissions. This helps keep our environment cleaner.
- **Independence:** Solar panels let you create your own electricity! This means less reliance on outside sources.

Disadvantages:

- **Initial Cost:** Yes, setting up solar panels can be pricey at first. But, over time, the savings can really add up!
- **Weather Dependence:** Solar power does rely on the weather. On cloudy days or at night, you'll get less energy.

Factors to Consider

- **Location:** Make sure your spot gets plenty of sunlight for the best performance.
- **Orientation:** Positioning matters. Proper direction and tilt angle will help capture the most sunlight.
- **System Size:** Figure out the right size based on how much space you have and how much energy you need.
- **Battery Storage:** Consider getting battery storage. It stores extra energy for nighttime or when it's cloudy.

Solar Power System Setup and Configuration

- **Solar Panels:** First off, you need to put those solar panels somewhere. Pick a solid, south-facing spot with little to no shade. This way, they'll soak up as much sun as possible. You can hook up the panels either in parallel or in series, depending on the voltage and current you need.
- **Inverter:** Next, let's talk about the inverter. You need to install a solar inverter, which changes the DC electricity from the panels into AC electricity, the type your home uses.
- **Battery Bank:** Now, onto the battery bank. Connect a battery bank to store extra solar energy. This ensures you've got power even when it's cloudy or dark outside. No one likes being left without electricity!
- **Monitoring System:** Last but not least, think about adding a monitoring system. This helps you keep an eye on how much energy you're making and using. It's a great way to ensure your solar power system runs smoothly and efficiently.

WIND ENERGY

Definition and Explanation

Wind energy is all about converting the wind's power into electricity using wind turbines. These are like big towers with huge blades that spin when the wind blows. When the blades spin, they transform the wind's kinetic energy into electrical energy. The spinning blades drive a generator, creating power that we can use to light our homes, workplaces, and other electrical devices.

Advantages and Disadvantages

Advantages:

- **So Much of It:** Wind energy is renewable, so there's always more of it. Because it's everywhere and doesn't run out, it's great for making power in a way that's friendly to our planet.
- **Super Clean:** Unlike burning fossil fuels, producing wind energy doesn't emit pollutants or greenhouse gases. This means it's good for the Earth and helps reduce carbon emissions.
- **Cheap to Keep Going:** Once a wind turbine is set up, it doesn't cost much to keep it running and can generate electricity for many years.

Disadvantages:

- **Hits and Misses:** Wind energy can't always be counted on since it depends on how fast and steady the wind blows. This makes the amount of electricity it generates fluctuate.
- **Eye-catching (Maybe Not in a Good Way):** Some people think wind turbines are visually unappealing. Because of that, they might not like having them around, leading to opposition and push-back against new projects in some areas.

Factors to Consider

Let's talk about wind energy for power generation. When you're thinking about this awesome energy source, keep these points in mind:
- **Wind Speed:** First off, check the wind speed where you are. You need to ensure the wind there is strong enough to make energy.
- **Site Selection:** Pick a spot without anything blocking the wind. You want a place where the wind can blow freely and minimize turbulence for your turbine to work efficiently.
- **Turbine Size:** Think about how big or small the turbine should be. This depends on how much energy you want, how windy it is, and how much space you've got.
- **Maintenance:** Keep an eye on the turbines. Regular maintenance is super important. This means checking them out regularly, adding oil when needed, and fixing any issues that pop up.

Wind Energy System Setup and Configuration

- **Turbine Installation:** Start by setting up the wind turbine on a strong tower that's tall enough to catch most of the wind. Make sure to attach the blades correctly to the hub and align everything just right for top performance.
- **Foundation:** Build a solid foundation since you want it to hold the tower's weight and not fall over in strong winds. A concrete pad or sturdy footing might be a good idea, depending on the type of ground and local regulations.
- **Electrical Connection:** Link up the generator with either the electrical grid or a battery storage system using the right wires and parts. You'll also need a wind turbine controller—this helps manage how your turbine runs and prevents it from overloading.
- **Safety Measures:** Safety first! Set up precautions to avoid any accidents while you're installing or running your turbine. Proper grounding is crucial. Add safety guards and put up warning signs so folks nearby know what's going on.

HYDROELECTRIC POWER

Definition and Explanation

Hydroelectric power, using the natural flow of water (like from rivers or dams), generates energy. It's a clever process: reservoirs or dams are built to raise water levels. This elevated water, driven by gravitational pull, spins turbines, which then power generators to create energy.

Advantages and Disadvantages

- **Renewable**: Hydroelectric power is amazing! It uses a constant stream of water to generate electricity without depleting precious resources. It's truly renewable!
- **Reliability**: Unlike wind and solar, hydroelectric power is always available—day or night, rain or shine. That's super dependable.
- **Low Emissions**: Hydroelectricity is great for our air and in fighting climate change because it barely releases any greenhouse gases.
- **Environmental Impact**: Large hydro projects can sometimes disrupt river ecosystems, alter natural landscapes, and displace wildlife and people.
- **High Initial Costs**: Building dams and hydro plants requires significant upfront investment and may involve complex engineering challenges.
- **Limited Location**: This power source is dependent on geography. Not every location has sufficient water or suitable land.

Factors to Think About

- **Water Availability:** Check how much water is there and if the rivers flow steadily. Do they change in different seasons? This helps keep power going strong.
- **Topography:** Look at the land. Is it good for building dams and reservoirs? It's important to see if the area fits hydroelectric plans.

- **Environmental Impact:** Think about what the project might do to nature. Could it change habitats or affect water quality and areas downstream?
- **Regulatory Compliance:** Make sure to follow local rules and laws about water use, protecting wildlife, and getting land permits.

Setting Up Hydroelectric Power Systems

- **Dam Construction:** Start by building a dam or reservoir. It stores water and creates a height difference that's good for generating energy. Make sure the dam can handle water pressure and has space for turbines.
- **Turbine Installation:** Place turbines inside the dam structure or at the bottom of the reservoir. These will capture the energy of the flowing water. Then, link the turbines to generators. This turns mechanical energy into electricity.
- **Transmission Infrastructure:** Set up transmission lines and electrical systems to send hydroelectric power to people who need it. Work with utility companies and regulatory agencies so everything fits into the grid.
- **Environmental Mitigation:** Take steps to lessen any bad effects on nature from hydroelectric projects. Use fish ladders, sediment traps, and work on restoring habitats to keep ecological balance intact.

BIOMASS AND GEOTHERMAL ENERGY

Biomass energy really shines as a green hero. It comes from things like wood, farm crops, and even leftover food waste. This energy is kind to our planet (and it doesn't run out)! It offers a clean, trusty option compared to regular energy sources.

Overview of Geothermal Power

Geothermal energy is another cool way we create power. It uses the Earth's natural heat—whether deep inside or near the top. This smart method taps into the planet's own warmth for power and heating or cooling. Best part? It's renewable, just like biomass!

Pros and Cons

- **Advantages:**
    - **Renewable:** Biomass and geothermal energy both keep coming back naturally. No worries about running out, which means less reliance on resources that will eventually deplete.
    - **Clean:** These energy types cut down on harmful gases (the ones that mess with our climate). So, they help keep our planet healthy and green.
    - **Reliable:** With biomass and geothermal systems, we get steady power. Even when the weather's not great, they are solid as a rock.
- **Disadvantages:**
    - **Resource Heavy:** For biomass energy, we need lots of land and water. Growing and processing these materials might take up space where we could grow food instead.
    - **Location-Based:** Geothermal systems need special geological spots to work right. Not every place has the right underground conditions to make it happen.

Factors to Think About

When you're thinking about different kinds of energy for making power, keep an eye on these things:

- **Biomass Energy:**
    - *Fuel Availability:* Check how easy it is to get fuels around you. Look at factors like land use, forestry practices, and waste management. This helps assess if the fuel is sustainable.
    - *Conversion Efficiency:* Look at how effectively the biomass is converted into energy. Doing this right means more power with less environmental impact.
- **Geothermal Energy:**
    - *Geological Conditions:* Conduct deep geological assessments to determine if your location can generate geothermal energy. Consider factors like heat flow, rock permeability, and hydrothermal activity.
    - *Heat Extraction Methods:* Explore different methods to extract heat, such as hydrothermal systems, enhanced geothermal systems, or geothermal heat pumps. These methods can optimize energy production with minimal complications.

System Setup & Configuration for Biomass and Geothermal Energy

Biomass Energy:

- **Fuel Preparation:** Gather your biomass fuels and get them ready. Make sure they're the right size, not too wet, and of good quality. This helps things burn better or convert more efficiently.
- **Combustion or Gasification Systems:** Now, set up those biomass systems! This includes things like boilers, furnaces, or gasifiers. They'll help turn all that organic stuff into heat or electricity.

Geothermal Energy:

- **Well Installation:** Time to drill geothermal wells! These wells let you reach underground heat. You'll need special drilling tools and techniques to ensure everything is done safely and works well.
- **Heat Extraction Systems:** Let's get those geothermal heat systems installed. You can use closed-loop or open-loop systems to move that underground heat to the surface. This heat can then be used for making electricity or for heating/cooling buildings.

Tailoring Your Energy System to Your Needs

- **Taking Stock of How Much Energy You Use:** Do a full check of what you use day-to-day. Look at every device and appliance that consumes electricity. This helps you figure out exactly what you need.
- **Doing Math for Power System Needs:** Figure out how much power you use. Check each gadget's wattage and how long they're on every day. Don't forget to consider the times when energy use is higher or lower.
- **Picking the Best Alternative Energy Source:** Think about factors like the resources you have, system efficiency, and environmental impact. Choose what fits best for your situation.
- **Why Preppers Must Plan for Energy:** It's crucial to plan ahead for energy needs in prepping and survival situations. Having a strong and lasting energy plan ensures you're ready and self-sufficient when emergencies happen.

PRACTICAL INSTANCES OF ALTERNATIVE ENERGY SYSTEMS

Off-Grid Homestead: We went into detail on this in Chapter 19, but it's one of the popular systems folks think about.

Urban Microgrid: Picture this—a bustling city, with a neighborhood that's turned its fate around, becoming a lively urban microgrid. Each building? Covered in rooftop solar panels, giving clean electricity to homes and businesses alike. And under the ground, geothermal heat pumps work using Earth's warmth to provide heating and cooling all year long. Thanks to teamwork and cool new tech, this urban microgrid shows how alternative energy can change cities. It reduces the need for centralized power and boosts community strength.

Rural Farmstead: Take a look at this family's journey to sustainable living on their rural farmstead. With big open fields and loads of sunlight, they've set up many solar panels powering their farmhouse and farm work. Plus, they've got plenty of biomass on their land. They use wood chips and farm waste to run a biogas generator for backup power on cloudy days. By planning well and being resourceful, they've created a self-sufficient haven that proves how versatile and reliable alternative energy can be in farming.

Off-Grid Cabin: Now, let's go deep into the forest with an outdoor lover who built his own off-grid cabin. Without regular utilities, he depends on solar power plus micro-hydroelectric energy from a nearby stream for all his energy needs. With his smart setup, he powers appliances, lights, and even communication gear! So he enjoys that rustic life without giving up modern stuff. This off-grid cabin shows what self-reliance and cleverness can do—letting people live off the land with minimal environmental impact.

TIPS AND TRICKS FOR SUCCESSFUL IMPLEMENTATION

- **Start Small, Scale Up:** First, take stock of what you need. Pick the best renewable energy source for your setup. Begin with a small system and grow it as you get more comfortable and confident.
- **Invest in Efficiency:** Why does this matter? Insulation, LED bulbs, and energy-saving gadgets make a big difference. Plus, smart tech can help you keep track of your energy use easily.
- **Embrace Redundancy:** Never put all your eggs in one basket! Use different renewable sources and have some backup power options. Consider generators or battery storage—just in case the lights go out.
- **Community Collaboration:** Don't do it alone. Connect with like-minded folks, join local energy groups, and take part in community projects. You'll learn from each other and share valuable resources.

PROJECT 31: CREATE A HYDROELECTRIC GENERATOR AT HOME

You can make a very basic hydroelectric generator that can be incredibly effective. It will take a good bit of time, energy, and patience though, so be prepared. The best way to do this is to get a kit offline. Alternatively, you can make your own from the core of a washing machine or parts from other appliances. It's up to you where you want to start and how much work you want to put into what can be a fairly long project.

Materials and Preparation

Strongly recommend a kit with a manual. It will ensure the best results. While you can make one from scratch or modify other items, one made from a kit with instructions will be more reliable in the event of a disaster or emergency.

Steps

1. Determine what your water source will be. You need something that moves, such as a creek, stream, or river. The more water that goes through, the more potential power you will generate.
2. Build the generator from the kit.
3. Test it using the source you plan to use.

A Functional Water Turbine

PROJECT 32: HOW TO CHARGE YOUR PHONE WHEN THERE IS NO ELECTRICITY

The last project is something that you can use even without an emergency. Phone batteries are notoriously short-lived, so now you can do something about it, as long as you have the right tools on hand.

For this one, you actually have a lot of choices. We are going to assume that the traditional methods are not available, such as a wall outlet, computer, and car. These are the ways you probably already charge your phone, so let's look at a few other ways you can get the job done without your usual energy supplies.

Steps

1. Have portable chargers. If you have a way to charge several of these at a time before leaving, this can be an easy way to charge your phone while moving between different locations without returning to a charging station.

Portable Charger

2. Use a solar charger. You can get one of these now, and have it when you need it. As long as there is enough sun to charge your phone, you can charge it from wherever you are.
3. Access a wood-fired charger. Yep, these actually have USB ports, so if you have a wood-fired charger, you have what you need to charge your phone.
4. Use a hand-crank charger. This one will take a good bit more effort, and it probably isn't going to buy you much charge, but if all you need is to get a quick communication out or check your map, it is all you need in a pinch.
5. Make a battery pack. Not efficient, but it can be effective if you have 8 D batters, some paper clips, and a tape.
   a. Attach the paper clips to the batteries' positive and negative terminals.
   b. Connect the charger at the end.

For the best ways to charge your phone, you should buy the tools you need now. Trying to build something from scraps or what you have on hand could work, but it requires as much luck as skill as you don't know what you will have on hand. It's best to be prepared and have a few alternative methods of charging your phone on hand in case something goes wrong.

Conclusions

As we reach the end of this comprehensive guide, it's important to take a moment to reflect on the journey we've embarked on together. This book has been crafted to equip you with the knowledge, skills, and confidence necessary to protect and sustain your family in any crisis. By delving into each of the chapters, you've not only learned practical techniques but have also embraced a mindset of resilience, adaptability, and independence.

The Path to Self-Sufficiency

The journey toward self-sufficiency is a dynamic and ongoing process. It requires continual learning, practice, and adaptation to ever-changing circumstances. Whether you're securing clean water, preserving food, mastering outdoor survival skills, or fortifying your home, each project and skill you've learned is a building block in creating a robust, self-reliant lifestyle. The knowledge you now possess is a powerful tool, not just for survival, but for thriving in the face of uncertainty.

Building a Resilient Community

While self-sufficiency starts with the individual, it is strengthened through community. The skills and projects outlined in this guide can be shared, taught, and practiced with others. By fostering a network of like-minded individuals, you can build a resilient community that supports and sustains each other in times of need. Remember, in a true crisis, collaboration and mutual aid often become as crucial as any individual skill.

Continual Preparedness

The world is unpredictable, and the skills you've gained are only as good as your commitment to maintaining and honing them. Regularly revisit the techniques and projects outlined in this book, practice them, and refine your methods. Stay informed about new developments in survival techniques, technology, and resources. Preparedness is not a one-time effort but a lifelong commitment to being ready for whatever comes your way.

Final Thoughts

As you close this book, know that you've taken a significant step toward ensuring the safety and well-being of yourself and your loved ones. The projects and skills you've mastered are more than just survival tactics; they are empowering tools that enable you to face the unknown with confidence.

Remember, the ultimate goal is not just to survive, but to live a life of independence, security, and peace of mind, no matter what challenges arise. The knowledge and skills you've acquired here are your insurance against uncertainty. With them, you are better prepared to face whatever the future holds, ensuring that you and your family not only survive but thrive in any crisis.

Stay vigilant, stay prepared, and always strive to be self-reliant. The future is uncertain, but with the right skills and mindset, you are ready to face it head-on.

Made in United States
Cleveland, OH
04 March 2025